Hands-On
Genetic Algorithms with Python

Apply genetic algorithms to solve real-world AI and machine learning problems

Eyal Wirsansky

Hands-On Genetic Algorithms with Python

Copyright © 2024 Packt Publishing

Group Product Manager: Ali Abidi
Publishing Product Manager: Tejashwini R
Book Project Manager: Shambhavi Mishra
Content Development Editor: Joseph Sunil
Technical Editor: Rahul Limbachiya
Copy Editor: Safis Editing
Proofreader: Joseph Sunil
Indexer: Rekha Nair
Production Designer: Aparna Bhagat
Senior DevRel Marketing Executive: Vinishka Kalra

First published: January 2020

Second Edition: June 2024

Production reference: 1140624

Published by Packt Publishing Ltd.
Grosvenor House
11 St Paul's Square
Birmingham
B3 1RB, UK.

ISBN 978-1-80512-379-8

www.packtpub.com

This book is dedicated to the loving memory of my parents, Gideon and Devorah Wirsansky. Your love and wisdom continue to guide and inspire me throughout my journey.

– Eyal Wirsansky

Contributors

About the author

Eyal Wirsansky is a senior data scientist, a seasoned software engineer, a leader in the technology community, and an artificial intelligence researcher.

In his graduate studies, he concentrated on genetic algorithms and neural networks, leading to the creation of an innovative supervised machine learning algorithm that blends these two approaches.

With a career spanning over 25 years, Eyal has pioneered developments in Voice over IP, healthcare, software development tools, and drone technology. He is currently a key member of the data platform team at Gradle, Inc., focusing on developing machine learning-driven features to boost developer productivity.

Beyond his professional endeavors, Eyal serves as an adjunct professor at Jacksonville University, where he teaches artificial intelligence. He leads the Jacksonville, Florida Java User Group and the AI for Enterprise virtual user group and is the author behind the developer-focused artificial intelligence blog, AI4Java.

About the reviewer

Ali Oztas is a data scientist with 5.5 years of experience mostly specializing in Natural Language Processing and Time Series. His masters' thesis was about runtime analysis of genetic algorithms, which is where he was introduced to the main topic of the book. Currently, he is working with Large Language Models and discovering new ways to use them.

Table of Contents

Part 2: Solving Problems with Genetic Algorithms

3

Using the DEAP Framework   45

4

Combinatorial Optimization   83

5

Constraint Satisfaction   117

6

Optimizing Continuous Functions   147

Part 3: Artificial Intelligence Applications of Genetic Algorithms

7

Enhancing Machine Learning Models Using Feature Selection   175

8

Hyperparameter Tuning of Machine Learning Models 195

9

Architecture Optimization of Deep Learning Networks 211

10

Reinforcement Learning with Genetic Algorithms 227

11

Natural Language Processing   247

12

Explainable AI, Causality, and Counterfactuals
with Genetic Algorithms   269

Part 4: Enhancing Performance with Concurrency and Cloud Strategies

13

14

Part 5: Related Technologies

15

16

Preface

Welcome to the exciting world of genetic algorithms and their application in **artificial intelligence** (**AI**), presented through the lens of Python programming. This book is a comprehensive guide that takes you from the fundamental principles of genetic algorithms to their powerful applications in AI, leveraging the practicality and simplicity of Python.

In the realm of computer science and problem-solving, genetic algorithms stand out for their unique approach to finding solutions. Mirroring the process of natural selection, these algorithms develop answers to problems in a way that's both fascinating and effective. Our journey begins with laying down the theoretical foundations of these algorithms, detailing core components and functionalities such as selection, crossover, and mutation. This sets the stage for advanced concepts and practical applications.

As we journey from theoretical foundations to practical implementations, we transition to tackling real-world problems using Python. These range from simple puzzles to complex optimization challenges. The focus then intensifies on AI applications, where genetic algorithms become pivotal tools in enhancing machine learning models, solving intricate reinforcement learning tasks, and delving into natural language processing and the emerging field of explainable AI.

Recognizing the importance of performance optimization in algorithmic applications, this book embarks on using concurrency and cloud computing to enhance the efficiency, speed, and scalability of genetic algorithms.

Our journey culminates in the fascinating realm of image reconstruction and other biologically inspired algorithms, revealing the unexpected and creative potential of genetic algorithms.

By the end of this journey, you will have gained a robust understanding of genetic algorithms, empowered with hands-on experience in applying them across various domains. This book is not just an academic exploration but also a practical guide that will prepare you to implement genetic algorithms in real-world scenarios effectively.

Whether you are a student, a professional in the field of AI, or just a curious mind eager to explore the fascinating intersection of biology and computing, this book promises to be a valuable resource. Join us as we unlock the potential of genetic algorithms in the dynamic field of AI.

Who this book is for

This book is for data scientists, software developers, and AI enthusiasts eager to break into genetic algorithms and apply them to real-world, intelligent applications as quickly as possible. It's crafted for those who wish to master genetic algorithms swiftly and effectively.

The primary audience for this book comprises three distinct groups:

- **Data scientists**: This book is a treasure trove for data scientists looking to integrate genetic algorithms into their toolbox. You will learn how to apply these algorithms to complex data problems, enhancing your predictive models and analytical capabilities.

- **Software developers**: As a software developer, this book serves as a gateway to integrating genetic algorithms into software solutions. Whether you're involved in developing AI applications, optimization tools, or complex system simulations, understanding how to implement and fine-tune genetic algorithms can significantly enhance the functionality and efficiency of your software. The book offers real-world examples and step-by-step guides, helping you to seamlessly incorporate these algorithms into your projects.

- **AI enthusiasts**: If you are passionate about AI and keen on expanding your knowledge in this domain, this book will be particularly beneficial. You will get a thorough grounding in genetic algorithms, an essential part of AI, and a hands-on approach to applying them in various AI-driven applications. The skills and knowledge acquired will be a significant stepping stone in your journey toward becoming an AI expert.

What this book covers

Chapter 1, An Introduction to Genetic Algorithms, introduces the basics and theory of genetic algorithms, drawing parallels with Darwinian evolution. It contrasts these algorithms with traditional ones, discussing their advantages, limitations, and practical applications. The chapter concludes by highlighting scenarios where genetic algorithms are particularly effective.

Chapter 2, Understanding the Key Components of Genetic Algorithms, provides a comprehensive exploration of genetic algorithms, starting with an outline of a genetic algorithm's basic flow. It then explores its core components and progresses to examine real coded genetic algorithms while covering advanced concepts such as elitism, niching, and sharing. It concludes with insights into the art of problem-solving using genetic algorithms.

Chapter 3, Using the DEAP Framework, introduces the DEAP framework (a versatile tool for solving real-world problems with genetic algorithms), guides you through its main modules, and demonstrates how to construct essential genetic algorithm components. This is demonstrated by programming a solution to the *OneMax* problem, followed by experiments with different settings of the genetic algorithm, revealing the impact of various modifications.

Chapter 4, Combinatorial Optimization, explores the application of genetic algorithms in combinatorial optimization while providing Python-based solutions using the DEAP framework. Key optimization problems covered include the *Knapsack*, *Traveling Salesman*, and *Vehicle Routing* problems. Additionally, the chapter discusses genotype-to-phenotype mapping and the balance between exploration and exploitation.

Chapter 5, Constraint Satisfaction, starts by defining the concept of constraint satisfaction and its relevance to search problems and combinatorial optimization. It then presents practical examples of these problems and their solutions using the DEAP framework. Key problems explored include the *N-Queen, nurse scheduling*, and *graph coloring* problems. Additionally, it explores the distinction between hard and soft constraints and their integration into the solution process.

Chapter 6, Optimizing Continuous Functions, explores the application of genetic algorithms in optimizing continuous search spaces, using real number-based genetic operators and DEAP framework tools. This chapter presents Python-based solutions for optimizing functions such as the *Eggholder*, *Himmelblau's*, and *Simionescu's* functions, incorporating techniques such as niching and sharing, and addressing constraints.

Chapter 7, Enhancing Machine Learning Models Using Feature Selection, explains the use of genetic algorithms to enhance supervised machine learning models through feature selection. It begins with an introduction to machine learning, focusing on regression and classification tasks, and discusses the benefits of feature selection in improving model performance. The chapter then demonstrates the use of genetic algorithms in identifying key features in a test regression problem and in optimizing a classification model using the *Zoo* dataset by isolating the most effective features.

Chapter 8, Hyperparameter Tuning of Machine Learning Models, explores the enhancement of supervised machine learning models through genetic algorithm-based hyperparameter tuning. After introducing hyperparameter tuning and the grid search concept in machine learning, it uses the *Wine* dataset and the adaptive boosting classifier as case studies. The chapter compares conventional grid search with a genetic algorithm-driven grid search for hyperparameter tuning, concluding with an attempt to further refine results using a direct genetic algorithm approach.

Chapter 9, Architecture Optimization of Deep Learning Networks, focuses on enhancing artificial neural network-based models through genetic algorithm-driven optimization of network architecture. It begins with an introduction to neural networks and deep learning, followed by a case study using the *Iris* dataset and the Multilayer Perceptron classifier. The chapter demonstrates network architecture optimization through a genetic algorithm-based solution and extends this approach to include simultaneous optimization of network architecture and model hyperparameters.

Chapter 10, Reinforcement Learning with Genetic Algorithms, illustrates the application of genetic algorithms in reinforcement learning, using two benchmark environments from the *Gymnasium* toolkit. It begins with an overview of reinforcement learning and an introduction to the Gymnasium toolkit and its Python interface. The focus then shifts to tackling the *MountainCar* and *CartPole* environments, developing genetic algorithm-based solutions for these specific challenges.

Chapter 11, Natural Language Processing, dives into the intersection of genetic algorithms and NLP. It introduces NLP and word embeddings and demonstrates their use in a *Semantle*-like mystery-word game where a genetic algorithm guesses the mystery word. The chapter also examines n-grams and document classification, employing genetic algorithms to select a concise and efficient feature subset, thereby enhancing understanding of the classifier's functionality.

Chapter 12, Explainable AI, Causality, and Counterfactuals with Genetic Algorithms, investigates the use of genetic algorithms for generating "what if" scenarios in explainable AI and causality, emphasizing counterfactual analysis. It introduces these fields and the concept of counterfactuals, followed by a practical application of the *German Credit Risk* dataset using genetic algorithms, uncovering valuable insights through counterfactual analysis.

Chapter 13, Accelerating Genetic Algorithms – the Power of Concurrency, explores enhancing genetic algorithms' performance using concurrency, focusing on multiprocessing. It discusses the benefits of applying concurrency and demonstrates it through Python's built-in functionalities as well as an external library. Various multiprocessing approaches are tested on a CPU-intensive variant of the *OneMax* problem, evaluating the performance improvements achieved.

Chapter 14, Beyond Local Resources – Scaling Genetic Algorithms in the Cloud, expands on enhancing genetic algorithm performance through a client-server model, using asynchronous I/O and cloud-based server computations on **AWS Lambda**. It discusses the split-architecture benefits, applies it to the *OneMax* problem, and guides through deploying a Flask server and an asyncio client, culminating in AWS Lambda deployment to showcase cloud-enhanced genetic algorithm efficiency.

Chapter 15, Evolutionary Image Reconstruction with Genetic Algorithms, explores genetic algorithms in image processing, focusing on reconstructing images with semi-transparent polygons. It starts with an overview of image processing in Python and explains creating images from scratch with polygons and calculating image differences. It concludes with developing a genetic algorithm-based program to reconstruct a famous painting using polygons, examining the evolutionary process and results.

Chapter 16, Other Evolutionary and Bio-Inspired Computation Techniques, introduces a variety of problem-solving and optimization techniques related to genetic algorithms. It covers genetic programming, **NeuroEvolution of Augmenting Topologies** (**NEAT**), and particle swarm optimization, demonstrating each through problem-solving Python programs. The chapter concludes with an overview of several other related computation paradigms.

To get the most out of this book

To maximize the benefits of this book, having a working knowledge of Python is essential. This prerequisite ensures that you can seamlessly understand and apply the concepts and examples provided. Whether you're looking to enhance your current role or transition into a new area of expertise, this book offers the practical knowledge and insights needed to succeed in the fascinating world of genetic algorithms.

Software/hardware covered in the book	Operating system requirements
Python 3.11	Windows, macOS, or Linux
DEAP 1.4.1	
Various Python libraries	
Amazon Web Services (Chapter 14)	

If you are using the digital version of this book, we advise you to type the code yourself or access the code from the book's GitHub repository (a link is available in the next section). Doing so will help you avoid any potential errors related to the copying and pasting of code.

Download the example code files

You can download the example code files for this book from GitHub at `https://github.com/PacktPublishing/Hands-On-Genetic-Algorithms-with-Python-Second-Edition`. If there's an update to the code, it will be updated in the GitHub repository.

We also have other code bundles from our rich catalog of books and videos available at `https://github.com/PacktPublishing/`. Check them out!

Conventions used

There are a number of text conventions used throughout this book.

`Code in text`: Indicates code words in text, database table names, folder names, filenames, file extensions, pathnames, dummy URLs, user input, and Twitter handles. Here is an example: "The server is built using Flask, while the client leverages Python's `asyncio` library for asynchronous operations."

A block of code is set as follows:

```
def busy_wait(duration):
    current_time = time.time()
    while (time.time() < current_time + duration):
        pass
```

When we wish to draw your attention to a particular part of a code block, the relevant lines or items are set in bold:

```
@app.route("/")
def welcome():
    return "<p>Welcome to our Fitness Evaluation Server!</p>"
```

Any command-line input or output is written as follows:

```
pip install Flask
```

Bold: Indicates a new term, an important word, or words that you see onscreen. For instance, words in menus or dialog boxes appear in **bold**. Here is an example: "Clicking on the function's name will bring us to the **Function overview** screen."

> **Tips or important notes**
> Appear like this.

Get in touch

Feedback from our readers is always welcome.

General feedback: If you have questions about any aspect of this book, email us at `customercare@ packtpub.com` and mention the book title in the subject of your message.

Errata: Although we have taken every care to ensure the accuracy of our content, mistakes do happen. If you have found a mistake in this book, we would be grateful if you would report this to us. Please visit `www.packtpub.com/support/errata` and fill in the form.

Piracy: If you come across any illegal copies of our works in any form on the internet, we would be grateful if you would provide us with the location address or website name. Please contact us at `copyright@packt.com` with a link to the material.

If you are interested in becoming an author: If there is a topic that you have expertise in and you are interested in either writing or contributing to a book, please visit `authors.packtpub.com`.

Share Your Thoughts

Once you've read *Hands-On Genetic Algorithms with Python, Second Edition*, we'd love to hear your thoughts! Scan the QR code below to go straight to the Amazon review page for this book and share your feedback.

https://packt.link/r/1-805-12379-3

Your review is important to us and the tech community and will help us make sure we're delivering excellent quality content.

Download a free PDF copy of this book

Thanks for purchasing this book!

Do you like to read on the go but are unable to carry your print books everywhere?

Is your eBook purchase not compatible with the device of your choice?

Don't worry, now with every Packt book you get a DRM-free PDF version of that book at no cost.

Read anywhere, any place, on any device. Search, copy, and paste code from your favorite technical books directly into your application.

The perks don't stop there, you can get exclusive access to discounts, newsletters, and great free content in your inbox daily

Follow these simple steps to get the benefits:

1. Scan the QR code or visit the link below

https://packt.link/free-ebook/978-1-80512-379-8

2. Submit your proof of purchase
3. That's it! We'll send your free PDF and other benefits to your email directly

Part 1:
The Basics of
Genetic Algorithms

In this section, you will be introduced to the key concepts of genetic algorithms, beginning with the Darwinian evolution analogy, basic principles, and theoretical foundations. We will then dive deeper into the components and implementation details of these algorithms, exploring their flow and various methods of selection, crossover, and mutation. The section also focuses on real-coded genetic algorithms and advanced concepts such as elitism, niching, and sharing, all setting the stage for problem-solving in subsequent sections.

This part contains the following chapters:

- *Chapter 1, An Introduction to Genetic Algorithms*
- *Chapter 2, Understanding the Key Components of Genetic Algorithms*

1

An Introduction to Genetic Algorithms

Drawing its inspiration from Charles Darwin's theory of natural evolution, one of the most fascinating techniques for problem-solving is the algorithm family suitably named evolutionary computation. Within this family, the most prominent and widely used branch is known as genetic algorithms. This chapter is the beginning of your journey to mastering this extremely powerful, yet extremely simple, technique.

In this chapter, we will introduce genetic algorithms and their analogy to Darwinian evolution before diving into their basic principles of operation and their underlying theory. We will then go over the differences between genetic algorithms and traditional ones and cover the advantages and limitations of genetic algorithms and their uses. We will conclude by reviewing cases where the use of a genetic algorithm may prove beneficial.

In this introductory chapter, we will cover the following topics:

- What are genetic algorithms?
- The theory behind genetic algorithms
- Differences between genetic algorithms and traditional algorithms
- Advantages and limitations of genetic algorithms
- When to use genetic algorithms

What are genetic algorithms?

Genetic algorithms are a family of search algorithms that are inspired by the principles of evolution in nature. By imitating the process of natural selection and reproduction, genetic algorithms can produce high-quality solutions for various problems involving search, optimization, and learning. At the same time, their analogy to natural evolution allows genetic algorithms to overcome some of the hurdles that are encountered by traditional search and optimization algorithms, especially for problems with a large number of parameters and complex mathematical representations.

In the rest of this section, we will review the basic ideas of genetic algorithms, as well as their analogy to the evolutionary processes transpiring in nature.

Darwinian evolution

Genetic algorithms implement a simplified version of the Darwinian evolution that takes place in nature. The principles of the Darwinian evolution theory can be summarized using the following principles:

- **The principle of variation**: The traits (attributes) of individual specimens belonging to a population may vary. As a result, the specimens differ from each other to some degree, for example, in their behavior or appearance.

- **The principle of inheritance**: Some traits are consistently passed on from specimens to their offspring. As a result, offspring resemble their parents more than they resemble unrelated specimens.

- **The principle of selection**: Populations typically struggle for resources within their given environment. The specimens possessing traits that are better adapted to the environment will be more successful at surviving and will also contribute more offspring to the next generation.

In other words, evolution maintains a population of individual specimens that vary from each other. Those who are better adapted to their environment have a greater chance of surviving, breeding, and passing their traits to the next generation. This way, as generations go by, species become more adapted to their environment and the challenges presented to them.

An important enabler of evolution is crossover or recombination – where offspring are created with a mix of their parents' traits. Crossover helps in maintaining the diversity of the population and in bringing together better traits over time. In addition, mutations – random variations in traits – can play a role in evolution by introducing changes that can result in a leap forward every once in a while.

The genetic algorithms analogy

Genetic algorithms seek to find the optimal solution for a given problem, whereas Darwinian evolution maintains a population of individual specimens. Genetic algorithms maintain a population of candidate solutions, called **individuals**, for that given problem. These candidate solutions are evaluated iteratively and used to create a new generation of solutions. Those who are better at solving this problem have a greater chance of being selected and passing their qualities to the next generation of candidate solutions. This way, as generations go by, candidate solutions get better at solving the problem at hand.

In the following sections, we will describe the various components of genetic algorithms that enable this analogy for Darwinian evolution.

Genotype

In nature, breeding, reproduction, and mutation are facilitated via the genotype – a collection of genes that are grouped into chromosomes. If two specimens breed to create offspring, each chromosome of the offspring will carry a mix of genes from both parents. Mimicking this concept, in the case of genetic algorithms, each individual is represented by a chromosome representing a collection of genes. For example, a chromosome can be expressed as a binary string, where each bit represents a single gene:

$$0\,1\,0\,1\,1\,1\,0\,1\,0$$

Figure 1.1: Simple binary-coded chromosome

Figure 1.1 shows an example of one such binary-coded chromosome, representing one particular individual.

Population

At any point in time, genetic algorithms maintain a population of individuals – a collection of candidate solutions for the problem at hand. Since each individual is represented by some chromosome, this population of individuals can be seen as a collection of such chromosomes:

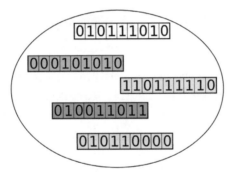

Figure 1.2: The population of individuals represented by binary-coded chromosomes

The population continually represents the current generation and evolves when the current generation is replaced by a new one.

Fitness function

At each iteration of the algorithm, the individuals are evaluated using a fitness function (also called the target function). This is the function we seek to optimize or the problem we are attempting to solve.

Individuals who achieve a better fitness score represent better solutions and are more likely to be chosen to reproduce and be represented in the next generation. Over time, the quality of the solutions improves, the fitness values increase, and the process can stop once a solution is found with a satisfactory fitness value.

Selection

After calculating the fitness of every individual in the population, a selection process is used to determine which of the individuals in the population will get to reproduce and create the offspring that will form the next generation.

This selection process is based on the fitness score of the individuals. Those with higher score values are more likely to be chosen and pass their genetic material to the next generation.

Individuals with low fitness values can still be chosen but with a lower probability. This way, their genetic material is not completely excluded, maintaining genetic diversity.

Crossover

To create a pair of new individuals, two parents are usually chosen from the current generation, and parts of their chromosomes are interchanged (crossed over) to create two new chromosomes representing the offspring. This operation is called crossover or recombination:

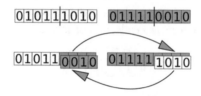

Figure 1.3: Crossover operation between two binary-coded chromosomes.
Source: `https://commons.wikimedia.org/wiki/File:Computational.`
`science.Genetic.algorithm.Crossover.One.Point.svg.`
Image by Yearofthedragon

Figure 1.3 illustrates a simple crossover operation of creating two offspring from two parents.

Mutation

The purpose of the mutation operator is to refresh the population, introduce new patterns into the chromosomes, and encourage search in uncharted areas of the solution space periodically and randomly.

A mutation may manifest itself as a random change in a gene. Mutations are implemented as random changes to one or more of the chromosome values; for example, flipping a bit in a binary string:

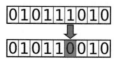

Figure 1.4: Mutation operator applied to a binary-coded chromosome

Figure 1.4 shows an example of the mutation operation.

Now, let's look at the theory behind genetic algorithms.

The theory behind genetic algorithms

The building-block hypothesis underlying genetic algorithms is that the optimal solution to the problem at hand is assembled of small building blocks, and as we bring more of these building blocks together, we get closer to this optimal solution.

Individuals in the population who contain some of the desired building blocks are identified by their superior scores. The repeated operations of selection and crossover result in better individuals conveying these building blocks to the next generations, while possibly combining them with other successful building blocks. This creates genetic pressure, thus guiding the population toward having more and more individuals with the building blocks that form the optimal solution.

As a result, each generation is better than the previous one and contains more individuals that are closer to the optimal solution.

For example, consider a population of four-digit binary strings where our goal is to find the string with the highest sum of digits. This is known as the **OneMax** problem and will be discussed in more detail later in this book. In this scenario, the digit 1 appearing at any of the four possible digit positions will be a good building block. As the algorithm progresses, it will identify solutions that have these building blocks and bring them together. Each generation will have more individuals with 1 value in various positions, ultimately leading to the string 1111, which incorporates all the desired building blocks. This process is illustrated in the following figure:

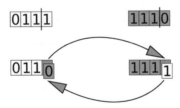

Figure 1.5: Demonstration of a crossover operation bringing the
building blocks of the optimal solution together

Figure 1.5 demonstrates how two individuals that are good solutions for this problem (each has three 1 values) create an offspring that is the best possible solution (four 1 bits – that is, the offspring on the right-hand side) when the crossover operation brings the desired building blocks of both parents together.

The schema theorem

A more formal expression of the building-block hypothesis is **Holland's schema theorem**, also called the **fundamental theorem of genetic algorithms**.

This theorem refers to schemata (the plural of schema), which are patterns (or templates) that can be found within chromosomes. Each schema represents a subset of chromosomes that have a certain similarity among them.

For example, if binary strings of length four represent the set of chromosomes, the schema *1*01* represents all those chromosomes that have a 1 in the leftmost position, 01 in the rightmost two positions, and either a 1 or a 0 in the second from the left position, since the * represents a **wildcard** value.

For each schema, we can assign two measurements:

- **Order**: The number of digits that are fixed (not wildcards)
- **Defining length**: The distance between the two furthermost fixed digits

The following table provides several examples of four-digit binary schemata and their measurements:

Schema	Order	Defining Length
1101	4	3
1*01	3	3
*101	3	2
*1*1	2	2
**01	2	1
1***	1	0
****	0	0

Table 1.1: Examples of four-digit binary schemata and their corresponding measurements

Each chromosome in the population corresponds to multiple schemata in the same way that a given string matches regular expressions. Chromosome 1101, for example, corresponds to every schemata that appears in this table since it matches each of the patterns they represent. If this chromosome has a higher score, it is more likely to survive the selection operation, along with all the schemata it represents. As this chromosome gets crossed over with another, or as it gets mutated, some of the schemata will survive and others will disappear. The schemata of low order and short defining length are the ones more likely to survive.

Consequentially, the schema theorem states that the frequency of schemata of low order, short defining length, and above-average fitness increases exponentially in frequency in successive generations. In other words, the smaller, simpler building blocks that represent the attributes that make a solution better will become increasingly present in the population as the genetic algorithm progresses. We will look at the difference between genetic and traditional algorithms in the next section.

Differences from traditional algorithms

There are several important differences between genetic algorithms and traditional search and optimization algorithms, such as gradient-based algorithms.

The key distinguishing factors are as follows:

- Maintaining a population of solutions
- Using a genetic representation of the solutions
- Utilizing the outcome of a fitness function
- Exhibiting a probabilistic behavior

We will describe these factors in greater detail in the following sections.

Population-based

The genetic search is conducted over a population of candidate solutions (individuals) rather than a single candidate. At any point in the search, the algorithm retains a set of individuals that form the current generation. Each iteration of the genetic algorithm creates the next generation of individuals.

In contrast, most other search algorithms maintain a single solution and iteratively modify it in search of the best solution. The gradient descent algorithm, for example, iteratively moves the current solution in the direction of the steepest descent, which is defined by the negative of the given function's gradient.

Genetic representation

Instead of operating directly on candidate solutions, genetic algorithms operate on their representations (or coding), often referred to as **chromosomes**. An example of a simple chromosome is a fixed-length binary string.

These chromosomes allow us to facilitate the genetic operations of crossover and mutation. Crossover is implemented by interchanging chromosome parts between two parents, while mutation is implemented by modifying parts of the chromosome.

A side effect of the use of genetic representation is decoupling the search from the original problem domain. Genetic algorithms are not aware of what the chromosomes represent and do not attempt to interpret them.

Fitness function

The fitness function represents the problem we would like to solve. The objective of genetic algorithms is to find the individuals that yield the highest score when this function is calculated for them.

Unlike many of the traditional search algorithms, genetic algorithms only consider the value that's obtained by the fitness function and do not rely on derivatives or any other information. This makes them suitable for handling functions that are hard or impossible to mathematically differentiate.

Probabilistic behavior

While many of the traditional algorithms are deterministic, the rules that are used by genetic algorithms to advance from one generation to the next are probabilistic.

For example, when selecting the individuals that will be used to create the next generation, the probability of selecting a given individual increases with the individual's fitness, but there is still a random element in making that choice. Individuals with low score values can still be chosen as well, although with a lower probability.

The mutation is probability-driven as well, usually occurs with low likelihood, and makes changes at one or more random location(s) in the chromosome.

The crossover operator can also have a probabilistic element. In some variations of genetic algorithms, the crossover will only occur at a certain probability. If no crossover takes place, both parents are duplicated into the next generation without change.

Despite the probabilistic nature of this process, the genetic-algorithm-based search is not random; instead, it uses the random aspect to direct the search toward areas in the search space where there is a better chance to improve the results. Now, let's look at the advantages of genetic algorithms.

Advantages of genetic algorithms

The unique characteristics of genetic algorithms that we discussed in the previous sections provide several advantages over traditional search algorithms.

The main advantages of genetic algorithms are as follows:

- Global optimization capability
- Can handle problems with a complex mathematical representation
- Can handle problems that lack mathematical representation
- Resilience to noise
- Support for parallelism and distributed processing
- Suitable for continuous learning

We will cover each of these in the upcoming sections.

Global optimization

In many cases, optimization problems have local maxima and minima points; these represent solutions that are better than those around them, but not the best overall.

The following figure illustrates the differences between global and local maximum and minimum points:

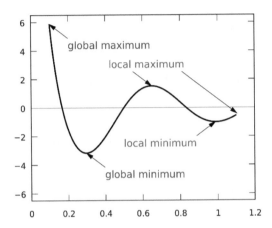

Figure 1.6: The global and local maxima and minima of a function.

Source: https://commons.wikimedia.org/wiki/File:Computational. science.Genetic.algorithm.Crossover.One.Point.svg.

Image by KSmrq

Most traditional search and optimization algorithms, and particularly those that are gradient-based, are prone to getting stuck in a local maximum rather than finding the global one. This is because, in the vicinity of a local maximum, any small change will degrade the score.

Genetic algorithms, on the other hand, are less sensitive to this phenomenon and are more likely to find the global maximum. This is due to the use of a population of candidate solutions rather than a single one, and the crossover and mutation operations that will, in many cases, result in candidate solutions that are distant from the previous ones. This is true so long as we manage to maintain the diversity of the population and avoid **premature convergence**, as we will mention in the next section.

Handling complex problems

Since genetic algorithms only require the outcome of the fitness function for each individual and are not concerned with other aspects of the fitness function, such as derivatives, they can be used for problems with complex mathematical representations or functions that are hard or impossible to differentiate.

Other complex cases where genetic algorithms excel include problems with a large number of parameters and problems with a mix of parameter types – for example, a combination of continuous and discrete parameters.

Handling a lack of mathematical representation

Genetic algorithms can be used for problems that lack mathematical representation altogether. One such case of particular interest is when the fitness score is based on human opinion. Imagine, for example, that we want to find the most attractive color palette to be used on a website. We can try different color combinations and ask users to rate the attractiveness of the site. We can apply genetic algorithms to search for the best scoring combination while using this opinion-based score as the fitness function's outcome. The genetic algorithm will operate as usual, even though the fitness function lacks any mathematical representation and there is no way to calculate the score directly from a given color combination.

As we will see in the next chapter, genetic algorithms can even deal with cases where the score of each individual cannot be obtained, so long as we have a way to compare two individuals and determine which of them is better. An example of this is a machine learning algorithm that drives a car in a simulated race. A genetic-algorithm-based search can optimize and tune the machine learning algorithm by having different versions of it compete against each other to determine which version is better.

Resilience to noise

Some problems present noisy behavior. This means that, even for similar input parameter values, the output value may be somewhat different every time it's measured. This can happen, for example, when the data that's being used is being read from sensor outputs, or in cases where the score is based on human opinion, as was discussed in the previous section.

While this kind of behavior can throw off many traditional search algorithms, genetic algorithms are generally resilient to it thanks to the repetitive operation of reassembling and reevaluating the individuals.

Parallelism

Genetic algorithms lend themselves well to parallelization and distributed processing. Fitness is calculated independently for each individual, which means all the individuals in the population can be evaluated concurrently.

In addition, the operations of selection, crossover, and mutation can each be performed concurrently on individuals and pairs of individuals in the population.

This makes genetic algorithms natural candidates for distributed as well as cloud-based implementation.

Continuous learning

In nature, evolution never stops. As the environmental conditions change, the population will adapt to them. Similarly, genetic algorithms can operate continuously in an ever-changing environment, and at any point in time, the best current solution can be fetched and used.

For this to be effective, the changes in the environment need to be slow concerning the generation turnaround rate of the genetic-algorithm-based search. Now that we've covered the advantages of genetic algorithms, let's look at the limitations.

Limitations of genetic algorithms

To get the most out of genetic algorithms, we need to be aware of their limitations and potential pitfalls.

The limitations of genetic algorithms are as follows:

- The need for special definitions
- The need for hyperparameter tuning
- Computationally intensive operations
- The risk of premature convergence
- No guaranteed solution

We will cover each of these in the upcoming sections.

Special definitions

When applying genetic algorithms to a given problem, we need to create a suitable representation for them – define the fitness function and the chromosome structure, as well as the selection, crossover, and mutation operators that will work for this problem. This can often prove to be challenging and time-consuming.

Luckily, genetic algorithms have already been applied to countless different types of problems, and many of these definitions have been standardized. This book covers numerous types of real-life problems and the way they can be solved using genetic algorithms. Use this as guidance whenever you are challenged by a new problem.

Hyperparameter tuning

The behavior of genetic algorithms is controlled by a set of hyperparameters, such as the population size and mutation rate. When applying genetic algorithms to the problem at hand, there are no exact rules for making these choices.

However, this is the case for virtually all search and optimization algorithms. After going over the examples in this book and doing some experimentation of your own, you will be able to make sensible choices for these values.

Computationally intensive

Operating on (potentially large) populations and the repetitive nature of genetic algorithms can be computationally intensive, as well as time-consuming, before a good result is reached.

These can be alleviated with a good choice of hyperparameters, implementing parallel processing, and in some cases, caching the intermediate results.

Premature convergence

If the fitness of one individual is much higher than the rest of the population, it may be duplicated enough that it takes over the entire population. This can lead to the genetic algorithm getting prematurely stuck in a local maximum, instead of finding the global one.

To prevent this from occurring, it is important to maintain the diversity of the population. Various ways to maintain diversity will be discussed in the next chapter.

No guaranteed solution

The use of genetic algorithms does not guarantee that the global maximum for the problem at hand will be found.

However, this is almost the case for any search and optimization algorithm, unless it is an analytical solution for a particular type of problem.

Generally, genetic algorithms, when used appropriately, are known to provide good solutions within a reasonable amount of time. Now, let's look at a few use cases for genetic algorithms.

Use cases for genetic algorithms

Based on the material we covered in the previous sections, genetic algorithms are best suited for the following types of problems:

- **Problems with complex mathematical representation**: Since genetic algorithms only require the outcome of the fitness function, they can be used for problems with target functions that are hard or impossible to differentiate (such as planning and scheduling), problems with a large number of parameters (such as image reconstruction), and problems with a mix of parameter types (such as hyperparameter optimization).

- **Problems with no mathematical representation**: Genetic algorithms don't require a mathematical representation of the problem, so long as a score value can be obtained, or a method is available to compare two solutions. This can be useful, for example, when solving reinforcement learning tasks or optimizing the architecture of a deep learning model.

- **Problems involving a noisy environment**: Genetic algorithms are resilient to conditions where data may not be consistent, such as information originating from sensor output or human-based scoring; for example, choosing the best color palette for a website based on customers' feedback and usage patterns.

- **Problems involving an environment that changes over time**: Genetic algorithms can respond to slow changes in the environment by continuously creating new generations that will adapt to these changes. Revisiting the website color palette example mentioned previously, the customers' favorite colors may change over time as per fashion trends. On the other hand, when a problem has a known and specialized way of being solved, using an existing traditional or analytic method is likely to be a more efficient choice.

With that, we've come to the end of this chapter.

Summary

In this chapter, we started by introducing genetic algorithms, their analogy to Darwinian evolution, and their basic principles of operation, including the use of population, genotype, the fitness function, and the genetic operators of selection, crossover, and mutation.

Then, we covered the theory underlying genetic algorithms by going over the building-block hypothesis and the schema theorem and illustrating how genetic algorithms work by bringing together superior, small building blocks to create the best solutions.

Next, we went over the differences between genetic algorithms and traditional ones, such as maintaining a population of solutions and using a genetic representation of those solutions.

We continued by covering the strengths of genetic algorithms, including their capacity for global optimization, handling problems with complex or non-existent mathematical representations, and resilience to noise, followed by their weaknesses, including the need for special definitions and hyperparameter tuning, as well as the risk of premature convergence.

We concluded by going over the cases where the use of a genetic algorithm may prove beneficial, such as in mathematically complex problems and optimization tasks in a noisy or ever-changing environment.

In the next chapter, we will delve deeper into the key components and the implementation details of genetic algorithms in preparation for the following chapters, where we will use them to code solutions for various types of problems.

Further reading

For more information on what we covered in this chapter, please refer to *Introduction to Genetic Algorithms*, from the book *Hands-On Artificial Intelligence for IoT*, by Amita Kapoor, January 2019, available at `https://subscription.packtpub.com/book/big_data_and_business_intelligence/9781788836067`.

2

Understanding the Key Components of Genetic Algorithms

In this chapter, we will dive deeper into the key components and the implementation details of genetic algorithms in preparation for the following chapters, where we will use genetic algorithms to create solutions for various types of problems.

First, we will outline the basic flow of a genetic algorithm, and then break it down into its different components while demonstrating various implementations of selection methods, crossover methods, and mutation methods. Next, we will look into real-coded genetic algorithms, which facilitate search in a continuous parameter space. This will be followed by an overview of the intriguing topics of elitism, niching, and sharing in genetic algorithms. Finally, we will study the art of solving problems using genetic algorithms.

By the end of this chapter, you will be able to do the following:

- Be familiar with the key components of genetic algorithms
- Understand the stages of the genetic algorithm flow
- Understand the genetic operators and become familiar with several of their variants
- Know the various options for stopping conditions
- Understand what modifications need to be made to the basic genetic algorithm when it's applied to real numbers
- Understand the mechanism of elitism
- Understand the concepts and implementation of niching and sharing
- Know the choices you have to make when you're starting to work on a new problem

The basic flow of a genetic algorithm

The main stages of the basic genetic algorithm flow are shown in the following flowchart:

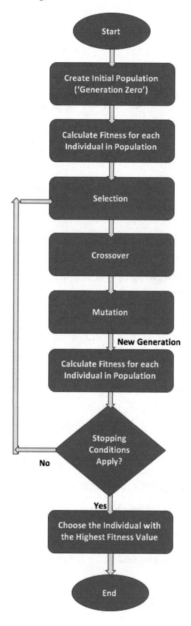

Figure 2.1: The basic flow of a genetic algorithm

These stages are described in detail in the following sections.

Creating the initial population

The initial population is a set of valid candidate solutions (individuals) that are chosen randomly. Since genetic algorithms use a chromosome to represent each individual, the initial population is a set of chromosomes. These chromosomes should conform to the chromosome format that we chose for the problem at hand – for example, binary strings of a certain length.

Calculating the fitness

The value of the fitness function is calculated for each individual. This is done once for the initial population, and then for every new generation after applying the genetic operators of selection, crossover, and mutation. As the fitness of each individual is independent of the others, this calculation can be done concurrently.

Since the selection stage that follows the fitness calculation usually considers individuals with higher fitness scores to be better solutions, genetic algorithms are naturally geared toward finding the maximum value(s) of the fitness function. If we have a problem where the minimum value is desired, the fitness calculation should inverse the original value – for example, by multiplying it by a value of (-1).

Applying selection, crossover, and mutation

Applying the genetic operators of selection, crossover, and mutation to the population results in the creation of a new generation that is based on better individuals than the current ones.

- The **selection** operator is responsible for selecting individuals from the current population in a way that gives an advantage to better individuals. Examples of selection operators are given in the *Selection methods* section.

- The **crossover** (or **recombination**) operator creates offspring from the selected individuals. This is usually done by taking two selected individuals at a time and interchanging parts of their chromosomes to create two new chromosomes representing the offspring. Examples of selection operators are given in the *Crossover methods* section.

- The **mutation** operator can randomly introduce a change to one or more of the chromosome values (genes) of each newly created individual. The mutation usually occurs with a very low probability. Examples of mutation operators are given in the *Mutation methods* section.

Checking the stopping conditions

There can be multiple conditions to check against when determining whether the process can stop. The two most commonly used stopping conditions are as follows:

- A maximum number of generations has been reached. This also serves to limit the runtime and computing resources that are consumed by the algorithm.

- There was no noticeable improvement over the last few generations. This can be implemented by storing the best fitness value that was achieved at every generation and comparing the current best value to the one that was achieved a predefined number of generations ago. If the difference is smaller than a certain threshold, the algorithm can stop.

Here are some other possible stopping conditions:

- The performance of the current best individual meets or exceeds the requirements of the specific use case

- A predetermined amount of time has elapsed since the process began

- A certain cost or budget has been consumed, such as CPU time and/or memory

- The best solution has taken over a portion of the population that is larger than a preset threshold

To summarize, the genetic algorithm flow starts with a population of randomly generated candidate solutions (individuals), which are evaluated against the fitness function. The heart of the flow is a loop where the genetic operators of selection, crossover, and mutation are applied successively, followed by re-evaluation of the individuals. The loop continues until a stopping condition is encountered, upon which the best individual of the existing population is selected as our solution. Now, let's look at selection methods.

Selection methods

Selection is used at the beginning of each cycle of the genetic algorithm flow to pick individuals from the current population that will be used as parents for the individuals of the next generation. The selection is probability-based, and the probability of an individual being picked is tied to its fitness value, in a way that it gives an advantage to individuals with higher fitness values.

The following sections describe some of the commonly used selection methods and their characteristics.

Roulette wheel selection

In the roulette wheel selection method, also known as **fitness proportionate selection** (**FPS**), the probability of selecting an individual is directly proportionate to its fitness value. This is comparable to using a roulette wheel in a casino and assigning each individual a portion of the wheel proportional to its fitness value. When the wheel is turned, the odds of each individual being selected are proportional to the size of the portion of the wheel that it occupies.

For example, suppose we have a population of six individuals with fitness values, as shown in the following table. The relative portion of the roulette wheel dedicated to each individual is calculated based on these fitness values:

Individual	Fitness	Relative Portion
A	8	7%
B	12	11%
C	27	24%
D	4	3%
E	45	40%
F	17	15%

Table 2.1: Fitness value table

The matching roulette wheel is depicted in the following diagram:

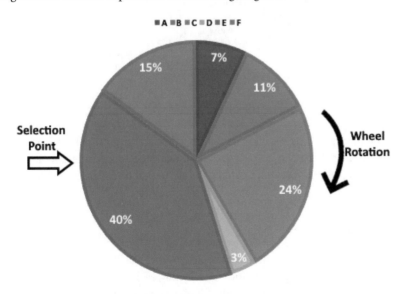

Figure 2.2: Roulette wheel selection example

Each time the wheel is turned, the selection point is used to choose a single individual from the entire population. The wheel is then turned again to select the next individual until we have enough individuals selected to fill the next generation. As a result, the same individual can be picked several times.

Stochastic universal sampling

Stochastic universal sampling (**SUS**) is a slightly modified version of the roulette wheel selection described previously. The same roulette wheel is used, with the same proportions, but instead of using a single selection point and turning the roulette wheel again and again until all needed individuals have been selected, we turn the wheel only once and use multiple selection points that are equally spaced around the wheel. This way, all the individuals are chosen at the same time, as depicted in the following diagram:

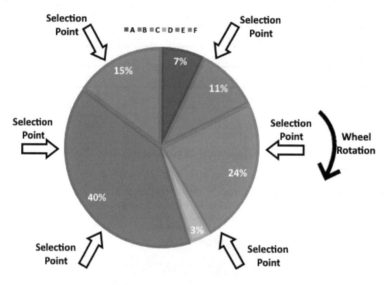

Figure 2.3: SUS example

This selection method prevents individuals with particularly high fitness values from saturating the next generation by getting chosen over and over again too many times. So, it provides weaker individuals with a chance to be chosen, reducing the somewhat unfair nature of the original roulette wheel selection method.

Rank-based selection

The rank-based selection method is similar to roulette wheel selection, but instead of directly using the fitness values to calculate the probabilities for selecting each individual, the fitness is used just to sort the individuals. Once sorted, each individual is given a rank representing their position, and the roulette probabilities are calculated based on these ranks.

For example, let's take the same population of six individuals we previously used with the same fitness values. To that, we'll add the rank of each individual. As the population size in our example is six, the highest-ranking individual gets a rank value of 6, the next one gets a rank value of 5, and so on. The relative portion of the roulette wheel dedicated to each individual is now calculated based on these rank values instead of using the fitness values:

Individual	Fitness	Rank	Relative Portion
A	8	2	9%
B	12	3	14%
C	27	5	24%
D	4	1	5%
E	45	6	29%
F	17	4	19%

Table 2.2: Fitness value table based on the relative portion

The matching roulette wheel is depicted in the following diagram:

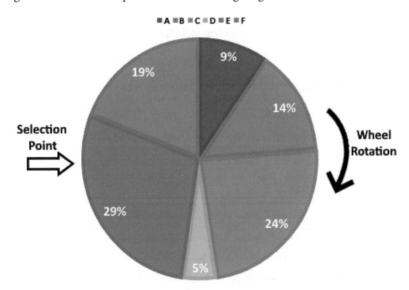

Figure 2.4: Rank-based selection example

Rank-based selection can be useful when a few individuals have much larger fitness values than all the rest. Using rank instead of raw fitness prevents these few individuals from taking over the entire population of the next generation as ranking eliminates the large differences.

Another useful case is when all individuals have similar fitness values. Here, rank-based selection will spread them apart, giving a clearer advantage to the better ones, even if the fitness differences are small.

Fitness scaling

While rank-based selection replaces each fitness value with the individual's rank, fitness scaling applies a scaling transformation to the raw fitness values and replaces them with the transformation's result. The transformation maps the raw fitness values into a desired range, as follows:

scaled fitness = a × (raw fitness) + b

Here, *a* and *b* are constants that we can select to achieve the desired range of the scaled fitness.

For example, if we use the same values from the previous examples, the range of the raw fitness values is between 4 (lowest fitness value, individual D) and 45 (highest fitness value, individual E). Suppose we want to map the values into a new range, between 50 and 100. We can calculate the values of the *a* and *b* constants using the following equations, representing these two individuals:

- $50 = a \times 4 + b$ (lowest fitness value)

- $100 = a \times 45 + b$ (highest fitness value)

Solving this simple system of linear equations will yield the following scaling parameter values:

a = 1.22, b = 45.12

This means that the scaled fitness values can be calculated as follows:

scaled fitness = 1.22 × (raw fitness) + 45.12

After adding a new column to the table containing the scaled fitness values, we can see that the range is indeed between 50 and 100, as desired:

Individual	Fitness	Scaled Fitness	Relative Portion
A	8	55	13%
B	12	60	15%
C	27	78	19%
D	4	50	12%
E	45	100	25%
F	17	66	16%

The matching roulette wheel is depicted in the following diagram:

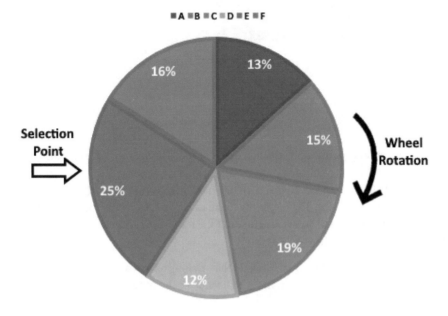

Figure 2.5: Roulette wheel selection example after fitness scaling

As *Figure 2.5* illustrates, scaling the fitness values to the new range provided a much more moderate partition of the roulette wheel compared to the original partition. The best individual (with a scaled fitness value of 100) is now only twice more likely to be selected than the worst one (with a scaled fitness value of 50), instead of being more than 11 times more likely to be chosen when using the raw fitness values.

Tournament selection

In each round of the tournament selection method, two or more individuals are randomly picked from the population, and the one with the highest fitness score wins and gets selected.

For example, suppose we have the same six individuals and the same fitness values we used in the previous examples. The following diagram illustrates randomly selecting three of them (A, B, and F), then announcing F as the winner since it has the largest fitness value (17) among these three individuals:

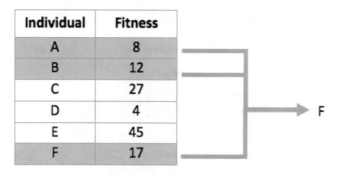

Figure 2.6: Tournament selection example with a tournament size of three

The number of individuals participating at each tournament selection round (three in our example) is suitably called **tournament size**. The larger the tournament size, the higher the chances that the best individuals will participate in the tournaments, and the lesser the chances of low-scoring individuals winning a tournament and getting selected.

An interesting aspect of this selection method is that, so long as we can compare any two individuals and determine which of them is better, the actual value of the fitness function is not needed. Next, we'll look at crossover methods.

Crossover methods

The crossover operator, also referred to as recombination, corresponds to the crossover that takes place during sexual reproduction in biology and is used to combine the genetic information of two individuals, serving as parents, to produce (usually two) offspring.

The crossover operator is typically applied with some (high) probability value. Whenever crossover is *not* applied, both parents are directly cloned into the next generation.

The following sections describe some of the commonly used crossover methods and their typical use cases. However, in certain situations, you may opt to use a problem-specific crossover method that will be more suitable for a particular case.

Single-point crossover

In the single-point crossover method, a location on the chromosomes of both parents is selected randomly. This location is referred to as the **crossover point** or **cut point**. Genes to the right of that point are swapped between the two parent chromosomes. As a result, we get two offspring, and each of them carries some genetic information from both parents.

The following diagram demonstrates a single-point crossover operation being conducted on a pair of binary chromosomes, with the crossover point located between the fifth and sixth genes:

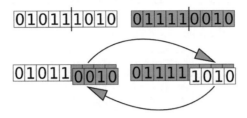

Figure 2.7: Single-point crossover example

Source: https://commons.wikimedia.org/wiki/File:Computational.
science.Genetic.algorithm.Crossover.One.Point.svg.

Image by Yearofthedragon.

In the next section, we will cover extensions of this method, namely two-point and k-point crossover.

Two-point and k-point crossover

In the two-point crossover method, two crossover points on the chromosomes of both parents are selected randomly. The genes residing between these points are swapped between the two parent chromosomes.

The following diagram demonstrates a two-point crossover carried out on a pair of binary chromosomes, with the first crossover point located between the third and fourth genes, and the other between the seventh and eighth genes:

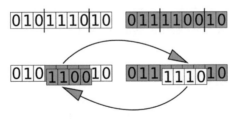

Figure 2.8: Two-point crossover example

Source: https://commons.wikimedia.org/wiki/File:Computational.
science.Genetic.algorithm.Crossover.Two.Point.svg.

Image by Yearofthedragon.

The two-point crossover method can be implemented by carrying out two single-point crossovers, each with a different crossover point. A generalization of this method is k-point crossover, where k represents a positive integer, and k crossover points are used.

Uniform crossover

In the uniform crossover method, each gene is independently determined by randomly choosing one of the parents. When the random distribution is 50%, each parent has the same chance of influencing the offspring, as illustrated in the following diagram:

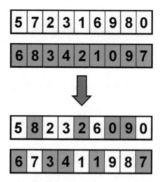

Figure 2.9: Uniform crossover example

Note that, in this example, the second offspring was created by complementing the choices made for the first offspring. However, both offspring can also be created independently of each other.

> **Important note**
> In this example, we used integer-based chromosomes, but it would work similarly with binary ones.

Since this method does not exchange entire segments of the chromosome, it has greater potential for diversity in the resulting offspring.

Crossover for ordered lists

In the previous example, we saw the results of a crossover operation on two integer-based chromosomes. While each of the parents had every value between 0 and 9 appear exactly once, each of the resulting offspring had certain values appearing more than once (for example, 2 in the top offspring and 1 in the other), and other values were missing (such as 4 in the top offspring and 5 in the other).

In some tasks, however, integer-based chromosomes may represent indices of an ordered list. For example, suppose we have several cities; we know the distance between each, and we need to find the shortest possible route through all of them. This is known as the traveling salesman problem and will be covered in detail in one of the following chapters.

If, for instance, we have four cities, a convenient way to represent a possible solution for this problem would be a four-integer chromosome showing the order of visiting the cities – for example, (1,2,3,4) or (3,4,2,1). A chromosome with two of the same values or missing one of the values, such as (1,2,2,4), will not represent a valid solution.

For such cases, alternative crossover methods were devised to ensure that the offspring that were created would still be valid. One of these methods, **ordered crossover (OX1)** will be covered in the following section.

OX1

The OX1 method strives to preserve the relative ordering of the parent's genes as much as possible. We will demonstrate it by using chromosomes with a length of six.

In this example, we used integer-based chromosomes, but it would work similarly with binary ones.

The first step is a two-point crossover with random cut points, as shown in the following diagram (with the parents depicted on the left-hand side):

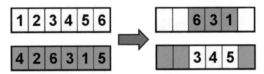

Figure 2.10: OX1 example – step 1

We will now start filling in the rest of the genes in each offspring by going over all the parent's genes in their original order, starting after the second cut point. For the first parent, we find a 6, but this is already present in the offspring, so we continue (with wrapping around) to *1*; this is already present too. The next in order is the *2*. Since *2* is not yet present in the offspring, we add it there, as shown in the following figure. For the second parent-offspring pair, we start with the parent's *5*, which is already present in the offspring, then move on to *4*, which is present as well, and end up with *2*, which is not present yet and therefore gets added. This is shown in the following diagram as well:

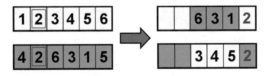

Figure 2.11: OX1 example – step 2

For the top parent, we now continue to 3 (already present in the offspring), and then 4, which gets added to the offspring. For the other parent, the next gene is 6. Since it's not present in the matching offspring, it gets added to it. The results are illustrated in the following diagram:

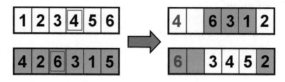

Figure 2.12: OX1 example – step 3

We continue similarly with the next genes not yet present in the offspring and fill in the last available spots, as depicted in the following diagram:

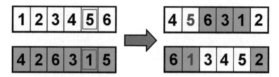

Figure 2.13: OX1 example – step 4

This completes the process of producing two valid offspring chromosomes, as the following diagram demonstrates:

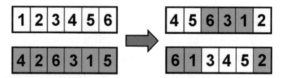

Figure 2.14: OX1 example – step 5

There are numerous other methods to implement crossover, some of which we will encounter later in this book. However, thanks to the versatility of genetic algorithms, you can always come up with your own methods. In the next section, we'll consider mutation methods.

Mutation methods

Mutation is the last genetic operator to be applied in the process of creating a new generation. The mutation operator is applied to the offspring that were created as a result of the selection and crossover operations.

The mutation operator is probability-based and usually occurs at a (very) low probability as it carries the risk of harming the performance of any individual it is applied to. In some versions of genetic algorithms, the mutation probability gradually increases as the generations advance to prevent stagnation and ensure the diversity of the population. On the other hand, if the mutation rate is increased excessively, the genetic algorithm will turn into the equivalent of a random search.

The following sections describe some of the commonly used mutation methods and their typical use cases. However, remember that you can always choose to use your own problem-specific mutation method that you deem more suitable for your particular use case.

Flip-bit mutation

When applying flip-bit mutation to a binary chromosome, one gene is randomly selected and its value is flipped (complemented), as shown in the following diagram:

Figure 2.15: Flip-bit mutation example

This can be extended to several random genes being flipped instead of just one.

Swap mutation

When applying the swap mutation to binary or integer-based chromosomes, two genes are randomly selected and their values are swapped, as shown in the following diagram:

Figure 2.16: Swap mutation example

This mutation operation is suitable for the chromosomes of ordered lists as the new chromosome still carries the same genes as the original one.

Inversion mutation

When applying the inversion mutation to binary or integer-based chromosomes, a random sequence of genes is selected and the order of the genes in that sequence is reversed, as depicted in the following diagram:

Figure 2.17: Inversion mutation example

Similar to the swap mutation, the inversion mutation operation is suitable for the chromosomes of ordered lists.

Scramble mutation

Another mutation operator that's suitable for the chromosomes of ordered lists is the scramble mutation. When applied, a random sequence of genes is selected and the order of the genes in that sequence is shuffled (or scrambled), as follows:

Figure 2.18: Scramble mutation example

In the next section, we will cover some other types of specialized operators that have been created for real-coded genetic algorithms.

Real-coded genetic algorithms

So far, we have seen chromosomes that represent binary or integer parameters. Consequently, the genetic operators were suitable for working on these types of chromosomes. However, we often encounter problems where the solution space is continuous. In other words, the individuals are made up of real (floating-point) numbers.

Historically, genetic algorithms used binary strings to represent integers as well as real numbers; however, this was not ideal. The precision of a real number represented using a binary string is limited by the length of the string (number of bits). Since we need to determine this length in advance, we may end up with binary strings that are too short, resulting in insufficient precision, or are overly long.

Moreover, when a binary string is used to represent a number, the significance of each bit varies by its location – the most significant bit being on the left. This can cause imbalance related to schemas – the patterns occurring in the chromosomes. For example, the schema 1**** (representing all five-digit binary strings starting with 1) and the schema ****1 (representing all five-digit binary strings ending with 1) both have an order of 1 and a defining length of 0; however, the first one carries much more significance than the other.

Instead of using binary strings, arrays of real-valued numbers were found to be a simpler and better approach. For example, if we have a problem involving three real-valued parameters, the chromosome will look like *[x1, x2, x3]*, where *x1*, *x2*, and *x3* represent real numbers, such as [1.23, 7.2134, -25.309] or [-30.10, 100.2, 42.424].

The various selection methods mentioned earlier in this chapter will work just the same for real-coded chromosomes as they only depend on the fitness of the individuals and not their representation.

However, the crossover and mutation methods we've covered so far will not be suitable for the real-coded chromosomes, so specialized ones need to be used. One important point to remember is that these crossover and mutation operations are applied separately for each dimension of the array that forms the real-coded chromosome. For example, if [1.23, 7.213, -25.39] and [-30.10, 100.2, 42.42] are parents that have been selected for the crossover operation, the crossover will be done separately for the following pairs:

- 1.23 and -30.10 (first dimension)
- 7.213 and 100.2 (second dimension)
- -25.39 and 42.42 (third dimension)

This is illustrated in the following diagram:

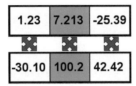

Figure 2.19: Real-coded chromosomes crossover example

Similarly, the mutation operator, when applied to a real-coded chromosome, will apply separately to each dimension.

Several of these real-coded crossover and mutation methods will be described in the following sections. Later, in *Chapter 6, Optimizing Continuous Functions*, we will get to see them in action.

Blend crossover

In the **blend crossover** (**BLX**) method, each offspring is randomly selected from the following interval created by its parents:

$$[parent_1 - \alpha(parent_2 - parent_1), parent_2 + \alpha(parent_2 - parent_1)]$$

The parameter, α, is a constant, whose value lies between 0 and 1. With larger values of α, the interval gets wider.

For example, if the parents' values are 1.33 and 5.72, the following will be the case:

- An α value of 0 will yield an interval of [1.33, 5.72] (similar to the interval between the parents)
- An α value of 0.5 will yield an interval of [-0.865, 7.915] (twice as wide as the interval between the parents)
- An α value of 1.0 will yield an interval of [-3.06, 10.11] (three times wider than the interval between the parents)

These examples are illustrated in the following diagram, where the parents are labeled as *p1* and *p2*, and the crossover interval is colored yellow:

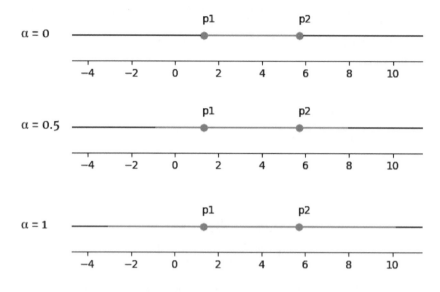

Figure 2.20: Blend crossover example

When using this crossover method, the α value is commonly set to 0.5.

Simulated binary crossover

The idea behind **simulated binary crossover** (**SBX**) is to imitate the properties of the single-point crossover that is commonly used with binary-coded chromosomes. One of these properties is that the average of the parents' values is equal to that of the offsprings' values.

When applying SBX, the two offspring are created from the two parents using the following formula:

$$offspring_1 = \frac{1}{2}\left[(1 + \beta)parent_1 + (1 - \beta)parent_2\right]$$

$$offspring_2 = \frac{1}{2}\left[(1 - \beta)parent_1 + (1 + \beta)parent_2\right]$$

Here, β is a random number referred to as the spread factor. This formula has the following notable properties:

- The average of the two offspring is equal to that of the parents, regardless of the value of β
- When the β value is 1, the offspring are duplicates of the parents
- When the β value is smaller than 1, the offspring are closer to each other than the parents were

- When the β value is larger than 1, the offspring are farther apart from each other than the parents were

For example, if the parents' values are 1.33 and 5.72, the following will be the case:

- A β value of 0.8 will yield 1.769 and 5.281 as offspring

- A β value of 1.0 will yield 1.33 and 5.72 as offspring

- A β value of 1.2 will yield 0.891 and 6.159 as offspring

These cases are illustrated in the following diagram, where the parents are labeled as *p1* and *p2* and the offspring are labeled as *o1* and *o2*:

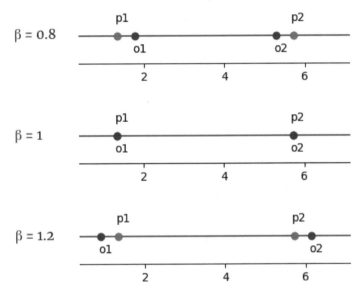

Figure 2.21: Simulated binary crossover example

In each of the preceding cases, the average value of the two offspring is 3.525, which is equal to the average value of the two parents.

Another property of binary single-point crossover that we would like to preserve is the similarity between offspring and parents. This translates to the random distribution of the β value. The probability of β should be much higher for values around 1, where the offspring are similar to the parents. To achieve that, the β value is calculated using another random value, denoted by u, that is uniformly distributed over the interval [0, 1]. Once the value of *u* is picked, β is calculated as follows:

If $u \leq 0.5$, we get the following: $\beta = (2u)^{\frac{1}{1+\eta}}$

Otherwise, we get the following: $\beta = \left[\frac{1}{2(1-u)}\right]^{\frac{1}{1+\eta}}$

Real mutation

One option for applying mutation in real-coded genetic algorithms is to replace any real value with a brand-new one, generated randomly. However, this can result in a mutated individual that has no relationship with the original individual.

Another approach is to generate a random real number that resides in the vicinity of the original individual. An example of such a method is the **normally distributed** (or **Gaussian**) **mutation**: a random number is generated using a normal distribution with a mean value of zero and some predetermined standard deviation, as shown in the following plot:

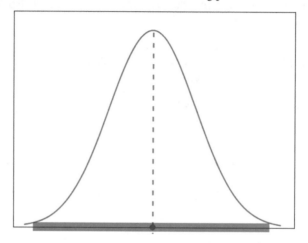

Figure 2.21: Gaussian mutation distribution example

In the next two sections, we will go over a couple of advanced topics, namely **elitism** and **niching**.

Understanding elitism

While the average fitness of the genetic algorithm population generally increases as generations go by, it is possible that, at any point, the best individual(s) of the current generation will be lost. This is due to the selection, crossover, and mutation operators altering the individuals in the process of creating the next generation. In many cases, the loss is temporary as these individuals (or better individuals) will be re-introduced into the population in a future generation.

However, if we want to guarantee that the best individual(s) always make it to the next generation, we can apply the optional elitism strategy. This means that the top *n* individuals (*n* being a small, predefined parameter) are duplicated into the next generation before we fill the rest of the available spots with offspring that are created using selection, crossover, and mutation. The elite individuals who were duplicated are still eligible for the selection process, so they can still be used as the parents of new individuals.

Elitism can sometimes have a significant positive impact on the algorithm's performance as it avoids the potential time waste needed for re-discovering good solutions that were lost in the genetic flow.

Another interesting way to enhance the results of genetic algorithms is the use of niching, as described in the next section.

Niching and sharing

In nature, all environments are further divided into multiple sub-environments, or niches, populated by various species taking advantage of the unique resources available in each niche, such as food and shelter. For example, a forest environment is comprised of treetops, shrubs, the forest floor, tree roots, and so on; each of these accommodates different species that are specialized for living in that niche and take advantage of its resources.

When several different species coexist in the same niche, they all compete over the same resources, and a tendency is created to search for new, unpopulated niches and populate them.

In the realm of genetic algorithms, this niching phenomenon can be used to maintain the diversity of the population, as well as to find several optimal solutions, each considered a niche.

For example, suppose our genetic algorithm seeks to maximize a fitness function that has several peaks of varying heights, such as the one in the following plot:

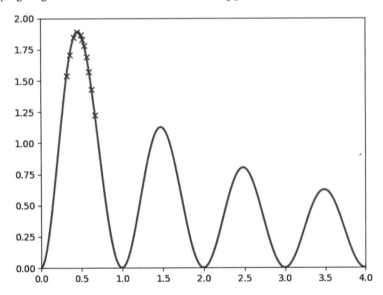

Figure 2.22: Expected genetic algorithm results without niching

As the genetic algorithm tends to find the global maximum, we expect, after a while, to see most of the population concentrating around the top peak. This is indicated in the preceding figure by the locations of the × marks on the function graph, representing individuals in the current generation.

However, there are implementations where, in addition to the global maximum, we would like to find some (or all) of the other peaks. To make this happen, we could think of each peak as a niche, offering resources in the amount proportional to its height. We then find a way to share (or divide) these resources among the individuals occupying them. This will ideally drive the population to be distributed accordingly, with the top peak attracting the most individuals as it offers the most reward, and the other peaks populated with decreasing portions of the population as they offer smaller amounts of reward.

This ideal situation is depicted in the following figure:

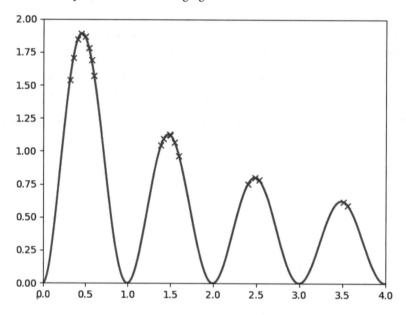

Figure 2.23: Ideal genetic algorithm results with niching

The challenge now is to implement this sharing mechanism. One option to accomplish sharing is to divide the raw fitness value of each individual with (some function of) the combined distances from all the other individuals. Another option would be to divide the raw fitness of each individual by the number of other individuals within a certain radius around it.

Serial niching versus parallel niching

Unfortunately, the niching concept, as described previously, can prove hard to implement as it increases the complexity of the fitness calculation. In practice, it will also require the population size to be the original one multiplied by the number of the expected peaks (which is generally unknown).

One way to overcome these issues is to find the peaks one at a time (serial niching) instead of attempting to find all of them at the same time (parallel niching). To implement serial niching, we use the genetic algorithm as usual and find the best solution. Then, we update the fitness function so that the area of the maximum point that was found is flattened, and repeat the process of the genetic algorithm.

Ideally, we will now find the next best peak as the original peak is no longer present. We can repeat this process iteratively and find the next best peak at each iteration.

The art of solving problems using genetic algorithms

Genetic algorithms provide us with a powerful and versatile tool that can be used to solve a wide array of problems and tasks. When we set to work on a new problem, we need to customize the tool and match it to that problem. This is done by making several choices, as described in the following paragraphs.

First, we need to determine the **fitness function**. This is how each individual will be evaluated, where larger values represent better individuals. The function does not have to be mathematical. It can be represented by an algorithm, a call to an external service, or even a result of a game played, to list a few options. We just need a way to programmatically retrieve the fitness value for any given proposed solution (individual).

Next, we need to choose an appropriate **chromosome encoding**. This is based on the parameters we send to the fitness function. So far, we have seen binary, integer, an ordered list, and real-coded examples. However, for some problems, we may need a mix of parameter types, or may even decide to create our own custom chromosome encoding.

Next, we need to pick a **selection** method. Most selection methods will work for any kind of chromosome type. If the fitness function is not directly accessible, but we still have a way to tell which of several candidate solutions is the best, we can consider utilizing the tournament selection method.

As we have seen in the preceding sections, the choice of **crossover** and **mutation** operators will be linked to the chromosome encoding of the individuals. Binary-coded chromosomes will have different crossover and mutation schemes than those that fit real-coded problems. Similar to the choice of chromosome encoding, here, too, you can design methods for crossover and mutation that fit your unique use case.

Lastly, there are the hyperparameters of the algorithm. The most common parameter values we need to set are as follows:

- Population size

- Crossover rate

- Mutation rate

- Max number of generations

- Other stopping condition(s)

- Elitism (used or not; what size)

For these parameters, we can choose what we deem as reasonable values and then tweak them, similar to how hyperparameters are dealt with in almost any other optimization and learning algorithm.

If making all these choices appears to be an overwhelming task, don't fret! In the chapters that follow, we will be iterating the process of making these choices time and again for the various types of problems we will tackle. After reading this book, you will be able to look at new problems and make your own wise choices.

Summary

In this chapter, you were introduced to the basic flow of the genetic algorithm. We then went over the key components of the flow, which included creating the population, calculating the fitness function, applying the genetic operators, and checking for stopping conditions.

Next, we went over various methods of selection, including roulette wheel selection, SUS, rank-based selection, fitness scaling, and tournament selection, and demonstrated the differences between them.

We continued by reviewing several crossover methods, including single-point, two-point, and k-point crossover, as well as OX1 and partially matched crossover.

You were then introduced to several mutation methods, including flip-bit mutation, followed by the swap, inversion, and scramble mutations.

Real-coded genetic algorithms were presented next, along with their specialized chromosome encoding and their custom genetic operators of crossover and mutation.

This was followed by an introduction to the concepts of elitism, niching, and sharing, as used in genetic algorithms.

In the last part of this chapter, you were presented with the various choices you need to make when approaching a problem to be solved using genetic algorithms, a procedure that will be repeated time and again throughout this book.

In the next chapter, the real fun begins – coding with Python! You will be introduced to DEAP, an evolutionary computation framework that can be used as a powerful tool for applying genetic algorithms to a wide array of tasks. DEAP will be used in the rest of this book as we develop Python programs that tackle numerous different challenges.

Further reading

For more information, please refer to *Chapter 8, Genetic Algorithms*, in the book *Artificial Intelligence with Python*, by Prateek Joshi, January 2017, available at `https://subscription.packtpub.com/book/big_data_and_business_intelligence/9781786464392/8`.

Part 2:
Solving Problems with
Genetic Algorithms

This part focuses on the application of genetic algorithms to various real-world problems using Python, starting with an exploration of the DEAP framework. We begin by tackling the foundational OneMax problem, showcasing the framework's capabilities. We then move on to more complex combinatorial optimization challenges, such as the traveling salesman and vehicle routing problems, and proceed with an in-depth look at constraint satisfaction problems, including the N-Queen and nurse scheduling problems. The part concludes by applying genetic algorithms to continuous search-space optimization, highlighting advanced techniques such as niching, sharing, and effective constraint management.

This part contains the following chapters:

- *Chapter 3, Using the DEAP Framework*
- *Chapter 4, Combinatorial Optimization*
- *Chapter 5, Constraint Satisfaction*
- *Chapter 6, Optimizing Continuous Functions*

3

Using the DEAP Framework

In this chapter – as promised – the real fun begins! You will be introduced to **Distributed Evolutionary Algorithms in Python** (**DEAP**) – a powerful and flexible evolutionary computation framework capable of solving real-life problems using genetic algorithms. After a brief introduction, you will get acquainted with two of its main modules – the creator and the toolbox – and learn how to create the various components needed for the genetic algorithm flow. We will then write a Python program that solves the OneMax problem – the Hello World of genetic algorithms – using the DEAP framework. This will be followed by a more concise version of the same program, where we'll take advantage of the built-in algorithms of the framework. We've saved the best for the last part of this chapter, where we will be experimenting with various settings of the genetic algorithm we created and discover the effects of our modifications.

By the end of this chapter, you will be able to do the following:

- Express your familiarity with the DEAP framework and its genetic algorithm modules
- Understand the concepts of the creator and toolbox modules in the DEAP framework
- Translate a simple problem into a genetic algorithm representation
- Create a genetic algorithm solution using the DEAP framework
- Understand how to use the DEAP framework's built-in algorithms to produce concise code
- Solve the OneMax problem using a genetic algorithm coded with the DEAP framework
- Experiment with various settings of the genetic algorithm and interpret the differences in the results

Technical requirements

Here are the technical requirements for this chapter.

> **Important note:**
> For the latest information regarding the technical requirements, please refer to the README file at: `https://github.com/PacktPublishing/Hands-On-Genetic-Algorithms-with-Python-Second-Edition/blob/main/README.md`

Python version

In this book, we will be using Python 3, version 3.11 or newer. Python can be downloaded from the Python Software Foundation at `https://www.python.org/downloads/`. Additional useful instructions can be found here: `https://realpython.com/installing-python/`.

Using a virtual environment

It is generally good practice to use a virtual environment when working on a Python-based project as it enables you to keep the dependencies of your project isolated from other Python projects, as well as the system's existing settings and dependencies.

One common way to create a virtual environment is by using **venv**, as described here: `https://docs.python.org/3/library/venv.html`.

Another popular way to manage Python environments and packages is using **conda**, as described here: `https://conda.io/projects/conda/en/latest/user-guide/tasks/manage-environments.html`.

> **Important note**
>
> When using a virtual environment, make sure you **activate** it before installing the required libraries, as described in the following section.

Installing the necessary libraries

Throughout this book, we will be using the DEAP library, as well as various other Python packages. There are a couple of options to install these dependencies, as outlined in the following subsections.

Using requirements.txt

Whether you choose to use a virtual environment or not, you can utilize the `requirements.txt` file we provide to install all the required dependencies at once. This file contains all the packages that will be used throughout this book and can be found in this book's GitHub repository at `https://github.com/PacktPublishing/Hands-On-Genetic-Algorithms-with-Python-Second-Edition/blob/main/requirements.txt`.

Typically, the `requirements.txt` file is used in conjunction with the `pip` utility and can be installed by applying the following command:

```
pip install -r /path/to/requirements.txt
```

Installing individual packages

If you prefer to install the required packages individually as you go through this book, the Technical requirements section of each chapter will mention the particular packages that will be used within that chapter.

To start with, we will need to install the DEAP library. The recommended ways to install DEAP are using `easy_install` or `pip`, like so:

```
pip install deap
```

For more information, check out the DEAP documentation: `https://deap.readthedocs.io/en/master/installation.html`.

If you prefer to install DEAP via Conda, consult the following link: `https://anaconda.org/conda-forge/deap`.

In addition, for this chapter, you will need the following packages:

- **NumPy**: `https://www.numpy.org/`
- **Matplotlib**: `https://matplotlib.org/`
- **Seaborn**: `https://seaborn.pydata.org/`

We are now ready to use DEAP. The framework's most useful tools and utilities will be covered in the next two sections. But first, we will get acquainted with DEAP and understand why we chose this framework for working with genetic algorithms.

The programs that will be used in this chapter can be found in this book's GitHub repository at `https://github.com/PacktPublishing/Hands-On-Genetic-Algorithms-with-Python-Second-Edition/tree/main/chapter_03`.

Check out the following video to see the Code in Action: `https://packt.link/OEBOd`.

Introduction to DEAP

As we have seen in the previous chapters, the basic ideas behind genetic algorithms and the genetic flow are relatively simple, and so are many of the genetic operators. Therefore, developing a program from scratch that implements a genetic algorithm to solve a particular problem is entirely feasible.

However, as is often the case when developing software, using a tried-and-true dedicated library or framework can make our lives easier. It helps us create solutions faster and with fewer bugs and gives us many options to choose from (and experiment with) right out of the box, without the need to reinvent the wheel.

Numerous Python frameworks have been created for working with genetic algorithms – PyGAD, GAFT, Pyevolve, and PyGMO, to mention a few. After looking into several options, we chose to use the DEAP framework for this book thanks to its ease of use and a large selection of features, as well as its extensibility and ample documentation.

DEAP is a Python framework that supports the rapid development of solutions using genetic algorithms, as well as other evolutionary computation techniques. DEAP offers various data structures and tools that prove essential when implementing a wide range of genetic-algorithm-based solutions.

DEAP has been developed at the Canadian Laval University since 2009 and is available under the GNU **Lesser General Public License (LGPL)**.

The source code for DEAP is available at `https://github.com/DEAP/deap` and the documentation can be found at `https://deap.readthedocs.io/en/master/`.

Using the creator module

The first powerful tool provided by the DEAP framework is the `creator` module. The `creator` module is used as a meta-factory, and it enables us to extend existing classes by augmenting them with new attributes.

For example, suppose we have a class called `Employee`. Using the `creator` tool, we can extend the `Employee` class by creating a `Developer` class, as follows:

```
from deap import creator
creator.create("Developer", Employee,\
               position="Developer", \
               programmingLanguages=set)
```

The first argument that's passed to the `create()` function is the desired name for the new class. The second argument is the existing base class to be extended. Then, each additional argument defines an attribute for the new class. If the argument is assigned a data structure (such as `dict` or `set`), it is added to the new class as an instance attribute that's initialized in the constructor. If the argument is a simple type, such as a literal, it's added as a class attribute that's shared among all instances of the class.

Consequently, the created `Developer` class will extend the `Employee` class and will have a class attribute, `position`, set to `Developer`, and an instance attribute, `programmingLanguages` of the `set` type, which is initialized in the constructor. So, effectively, the new class is equivalent to the following:

```
class Developer(Employee):
    position = "Developer"

    def __init__(self):
        self.programmingLanguages = set()
```

> **Important notes**
> 1. This new class exists within the `creator` module and therefore needs to be referenced as `creator.Developer`.
> 2. Extending the `numpy.ndarray` class is a special case that will be discussed later in this book.

When using DEAP, the `creator` module usually serves to create the `Fitness` class, as well as the `Individual` class, to be used by the genetic algorithm, as we will see next.

Creating the Fitness class

When using DEAP, fitness values are encapsulated within a `Fitness` class. DEAP enables fitness to be combined into several components (also called objectives), each having its own weight. The combination of these weights defines the behavior or strategy of the fitness for the given problem.

Defining the fitness strategy

To help define this strategy, DEAP comes with the abstract `base.Fitness` class, which contains a `weights` tuple. This tuple needs to be assigned values to define the strategy and make the class usable. This can be done by extending the base `Fitness` class using `creator`, in a similar manner to what we did with the preceding `Developer` class:

```
creator.create("FitnessMax", base.Fitness, weights=(1.0,))
```

This will yield a `creator.FitnessMax` class that extends the `base.Fitness` class, with the `weights` class attribute initialized to a value of `(1.0,)`.

> **Important note**
> Note the trailing comma in the `weights` definition when a single weight is defined. The comma is required because `weights` is a **tuple**.

The strategy of this `FitnessMax` class is to *maximize* the fitness values of the single-objective solutions throughout the genetic algorithm. Conversely, if we have a single-objective problem where we need to find a solution that *minimizes* the fitness value, we can use the following definition to create the appropriate minimizing strategy:

```
creator.create("FitnessMin", base.Fitness, weights=(-1.0,))
```

We can also define a class with a strategy for optimizing more than one objective, and with varying degrees of importance:

```
creator.create("FitnessCompound", base.Fitness,
    weights=(1.0, 0.2, -0.5))
```

This will create the `creator.FitnessCompound` class, which will utilize three different fitness components. The first will be given a weight of `1.0`, the second `0.2`, and the third `-0.5`. This fitness strategy will tend to maximize the first and the second components (or objectives) and minimize the third. In terms of importance, the first component has the most importance, followed by the third component and then the second one.

Storing the fitness values

While the `weights` tuple defines the fitness strategy, a matching tuple, called `values`, is used to contain the actual fitness values within the `base.Fitness` class. These values are obtained from a separately defined function, typically called `evaluate()`, as will be described later in this chapter. Just like the `weights` tuple, the `values` tuple contains one value for each fitness component (objective).

A third tuple, `wvalues`, contains the weighted values that are obtained by multiplying each component of the values tuple with its matching component of the `weights` tuple. Whenever the fitness values of an instance are set, the weighted values are calculated and inserted into `wvalues`. These are used internally for comparison operations between individuals.

The weighted fitness values may be lexicographically compared using the following operators:

```
>, <, >=, <=, ==, !=
```

Once the `Fitness` class is created, we can use it in the definition of the `Individual` class, as shown in the next subsection.

Creating the Individual class

The second common use of the `creator` tool in DEAP is defining the individuals that form the population for the genetic algorithm. As we saw in the previous chapters, the individuals in genetic algorithms are represented using a chromosome that can be manipulated by genetic operators. In DEAP, the `Individual` class is created by extending a base class that represents the chromosome. In addition, each instance in DEAP needs to contain its fitness function as an attribute.

To fulfill these two requirements, we can utilize `creator` to create the `creator.Individual` class, as shown in this example:

```
creator.create("Individual", list, \
               fitness=creator.FitnessMax)
```

This line provides the following two effects:

- The created `Individual` class extends the Python `list` class. This means that the chromosome that's used is of the `list` type.

- Each instance of this `Individual` class will have an attribute called `fitness`, of the `FitnessMax` class, which we created previously.

We will learn to use the `Toolbox` class in the next section.

Using the Toolbox class

The second mechanism offered by the DEAP framework is the `base.Toolbox` class. `Toolbox` is used as a container for functions (or operators) and enables us to create new operators by aliasing and customizing existing functions.

For example, suppose we have a function, `sumOfTwo()`, defined as follows:

```
def sumOfTwo(a, b):
    return a + b
```

Using `toolbox`, we can now create a new operator, `incrementByFive()`, which customizes the `sumOfTwo()` function, as follows:

```
from deap import base
toolbox= base.Toolbox()
toolbox.register("incrementByFive", sumOfTwo, b=5)
```

The first argument that's passed to the `register()` toolbox function is the desired name (or alias) for the new operator. The second argument is the existing function to be customized. Then, each additional (optional) argument is automatically passed to the customized function whenever we call the new operator. For example, look at this definition:

```
toolbox.incrementByFive(10)
```

Calling the preceding function is equivalent to calling this:

```
sumOfTwo(10, 5)
```

This is because the b argument has been fixed to a value of 5 by the definition of the `incrementByFive` operator.

Creating genetic operators

In many cases, the `Toolbox` class is used to customize existing functions from the `tools` module. The `tools` module contains numerous handy functions related to the genetic operations of *selection*, *crossover*, and *mutation*, as well as initialization utilities.

For example, the following code defines three aliases that will be later used as genetic operators:

```
from deap import tools
toolbox.register("select",tools.selTournament,tournsize=3)
toolbox.register("mate", tools.cxTwoPoint)
toolbox.register("mutate", tools.mutFlipBit, indpb=0.02)
```

The three aliases are defined as follows:

- `select` is registered as an alias to the existing `tools` function, `selTournament()`, with the `tournsize` argument set to 3. This creates a `toolbox.select` operator that performs *tournament selection* with a tournament size of 3.

- `mate` is registered as an alias to the existing `tools` function, `cxTwoPoint()`. This results in a `toolbox.mate` operator that performs *two-point crossover*.

- `mutate` is registered as an alias to the existing `tools` function, `mutFlipBit`, with the `indpb` argument set to `0.02`, providing a `toolbox.mutate` operator that performs *flip-bit mutation* with 0.02 as the probability for each attribute to be flipped.

The `tools` module provides implementations of various genetic operators, including several of the ones we mentioned in the previous chapter.

Selection functions can be found in the `selection.py` file. Some of them are as follows:

- `selRoulette()` implements **roulette wheel selection**

- `selStochasticUniversalSampling()` implements **Stochastic Universal Sampling (SUS)**

- `selTournament()` implements **tournament selection**

Crossover functions can be found in the `crossover.py` file:

- `cxOnePoint()` implements **single-point crossover**

- `cxUniform()` implements **uniform crossover**

- `cxOrdered()` implements **ordered crossover (OX1)**

- `cxPartialyMatched()` implements **partially matched crossover (PMX)**

A couple of the **Mutation** functions that can be found in the `mutation.py` file are as follows:

- `mutFlipBit()` implements **flip-bit mutation**

- `mutGaussian()` implements **normally distributed mutation**

Creating the population

The `init.py` file of the `tools` module contains several functions that can be useful for creating and initializing the population for the genetic algorithm. One particularly useful function is `initRepeat()`, which accepts three arguments:

- The container type in which we would like to put the resulting objects

- The function that's used to generate objects that will be put into the container

- The number of objects we want to generate

For example, the following line of code will produce a list of 30 random numbers between 0 and 1:

```
randomList = tools.initRepeat(list, random.random, 30)
```

In this example, `list` is the type serving as the container to be filled, `random.random` is the generator function, and `30` is the number of times we will call the function to generate values that fill the container.

What if we wanted to fill the list with integer random numbers that are *either* 0 or 1? We could, for example, create a function that utilizes `random.radint()` to generate a single random value of 0 or 1, and then use it as the generator function of `initRepeat()`, as shown in the following code snippet:

```
def zeroOrOne():
    return random.randint(0, 1)

randomList = tools.initRepeat(list, zeroOrOne, 30)
```

Alternatively, we can take advantage of `toolbox`, as follows:

```
toolbox.register("zeroOrOne", random.randint, 0, 1)
randomList = tools.initRepeat(list, toolbox.zeroOrOne, 30)
```

Here, instead of explicitly defining the `zeroOrOne()` function, we created the `zeroOrOne` operator (or alias), which calls `random.radint()` with the fixed parameters of 0 and 1.

Calculating the fitness

As mentioned previously, while the `Fitness` class defines the fitness *weights* that determine its strategy (such as *maximization* or *minimization*), the actual fitness values are obtained from a separately defined function. This fitness calculation function is typically registered with the `toolbox` module using an alias of `evaluate`, as shown in the following code snippet:

```
def someFitnessCalculationFunction(individual):
    return _some_calculation_of_the_fitness

toolbox.register("evaluate",someFitnessCalculationFunction)
```

In this example, `someFitnessCalculationFunction()` calculates the fitness for any given individual, while `evaluate` is registered as its alias.

We are finally ready to put our knowledge to use and solve our first problem using a genetic algorithm written with DEAP. We'll do this in the next section.

The OneMax problem

The OneMax (or One-Max) problem is a simple optimization task that is often used as the *Hello World* of genetic algorithm frameworks. We will use this problem for the rest of this chapter to demonstrate how DEAP can be used to implement a genetic algorithm.

The OneMax task is to find the binary string of a given length that maximizes the sum of its digits. For example, the OneMax problem of length 5 will consider candidates such as the following:

- 10010 (sum of digits = 2)
- 01110 (sum of digits = 3)
- 11111 (sum of digits = 5)

Obviously (to us), the solution to this problem is always the string that comprises all 1s. However, the genetic algorithm does not have this knowledge and needs to blindly look for the solution using its genetic operators. If the algorithm does its job, it will find this solution, or at least one close to it, within a reasonable amount of time.

The DEAP framework's documentation uses the OneMax problem as its introductory example (`https://github.com/DEAP/deap/blob/master/examples/ga/onemax.py`). In the following sections, we will describe our version of DEAP's OneMax example.

Solving the OneMax problem with DEAP

In the previous chapter, we mentioned several choices that need to be made when solving a problem using the genetic algorithm approach. As we tackle the OneMax problem, we will make these choices in a series of steps. In the chapters to follow, we will keep using the same series of steps as we apply the genetic algorithms approach to various types of problems.

Choosing the chromosome

Since the OneMax problem deals with binary strings, the choice of chromosome is easy – each individual will be represented with a binary string that directly represents a candidate solution. In the actual Python implementation, this will be implemented as a list containing integer values of either 0 or 1. The length of the chromosome matches the size of the OneMax problem. For example, for a OneMax problem of size 5, the 10010 individual will be represented by `[1, 0, 0, 1, 0]`.

Calculating the fitness

Since we want to find the individual with the largest sum of digits, we are going to use the `FitnessMax` strategy. As each individual is represented by a list of integer values of either 0 or 1, the fitness value will be directly calculated as the sum of the elements in the list – for example, `sum([1, 0, 0, 1, 0]) = 2`.

Choosing the genetic operators

Now, we need to decide on the genetic operators to be used – *selection*, *crossover*, and *mutation*. In the previous chapter, we examined several different types of each of these operators. Choosing these genetic operators is not an exact science, and we can usually experiment with several choices. But while selection operators can typically work with any chromosome type, the crossover and mutation operators we choose need to match the chromosome type we use; otherwise, they could produce invalid chromosomes.

For the selection operator, we can start with *tournament* selection because it is simple and efficient. Later, we can experiment with other selection strategies, such as *roulette wheel* selection and *SUS*.

For the *crossover* operator, either the *single-point* or *two-point* crossover operator will be suitable as the result of crossing over two binary strings using these methods will produce a valid binary string.

The *mutation* operator can be the simple *flip-bit* mutation, which works well for binary strings.

Setting the stopping condition

It is always a good idea to put a limit on the number of generations to guarantee that the algorithm doesn't run forever. This gives us one stopping condition.

In addition, since we happen to know the best solution for the OneMax problem – a binary string with all 1s, and a fitness value equal to the length of the individual – we can use that as a second stopping condition.

> **Important note**
> For a real-world problem, we typically don't have this kind of knowledge in advance.

If either of these conditions is met – that is, the number of generations reaches the limit *or* the best solution is found – the genetic algorithm will stop.

Implementing with DEAP

Putting it all together, we can finally start coding our solution to the OneMax problem using the DEAP framework.

The complete program containing the code snippets shown in this section can be found here: `https://github.com/PacktPublishing/Hands-On-Genetic-Algorithms-with-Python-Second-Edition/blob/main/chapter_03/01_OneMax_long.py`.

Setting up

Before we start the actual genetic algorithm flow, we need to set things up. The DEAP framework has quite a distinct way of doing this, as shown in the rest of this section:

1. We start by importing the essential modules of the DEAP framework, followed by a couple of useful utilities:

    ```
    from deap import base
    from deap import creator
    from deap import tools
    import random
    import matplotlib.pyplot as plt
    ```

2. Next, we must declare a few constants that set the parameters for the problem and control the behavior of the genetic algorithm:

    ```
    # problem constants:
    ONE_MAX_LENGTH = 100    # length of bit string to be
                            # optimized


    # Genetic Algorithm constants:
    POPULATION_SIZE = 200 # number of individuals in
                          # population
    P_CROSSOVER = 0.9       # probability for crossover
    P_MUTATION = 0.1        # probability for mutating
                            # an individual
    MAX_GENERATIONS = 50  # max number of generations for
                          # stopping condition
    ```

3. One important aspect of the genetic algorithm is the use of probability, which introduces a random element to the behavior of the algorithm. However, when experimenting with the code, we may want to be able to run the same experiment several times and get repeatable results. To accomplish this, we must set the random `seed` function to a constant number of some value, as shown in the following code snippet:

    ```
    RANDOM_SEED = 42
    random.seed(RANDOM_SEED)
    ```

> **Tip**
>
> At some point, you may decide to remove these lines of code, so separate runs could produce somewhat different results.

4. As we saw earlier in this chapter, the `Toolbox` class is one of the main utilities provided by the DEAP framework, enabling us to register new functions (or operators) that customize existing functions using pre-set arguments. Here, we'll use it to create the `zeroOrOne` operator, which customizes the `random.randomint(a, b)` function. This function normally returns a random integer, N, such that a ≤ N ≤ b. By fixing the two arguments, a and b, to the values 0 and 1, the `zeroOrOne` operator will randomly return either 0 or 1 when it's called later in the code. The following code snippet defines the `toolbox` variable, and then uses it to register the `zeroOrOne` operator:

```
toolbox = base.Toolbox()
toolbox.register("zeroOrOne", random.randint, 0, 1)
```

5. Next, we need to create the `Fitness` class. Since we only have one objective here – the sum of digits – and our goal is to maximize it, we'll choose the `FitnessMax` strategy and use a `weights` tuple with a single positive weight, as shown in the following code snippet:

```
creator.create("FitnessMax", base.Fitness, \
                weights=1.0,))
```

6. in DEAP, the convention is to use a class called `Individual` to represent each of the population's individuals. This class is created with the help of the `creator` tool. In our case, `list` serves as the base class, which is used as the individual's chromosome. The class is augmented with the `fitness` attribute, initialized to the `FitnessMax` class that we defined earlier:

```
creator.create("Individual", list, \
                fitness=creator.FitnessMax)
```

7. Next, we must register the `individualCreator` operator, which creates an instance of the `Individual` class that's filled with random values of 0 or 1. This is done by utilizing the previously defined `zeroOrOne` operator. This definition makes use of the `initRepeat` operator that was mentioned earlier as the base class. It can be customized using the following arguments:

    • The `Individual` class can be used as the container type in which the resulting objects will be placed

    • The `zeroOrOne` operator can be used as the function to generate objects

    • The `ONE_MAX_LENGTH` constant can be used as the number of objects we want to generate (currently set to `100`)

    Since the objects that are generated by the `zeroOrOne` operator are integers with random values of 0 or 1, the resulting `individualCreator` operator will fill an `Individual` instance with 100 randomly generated values of 0 or 1:

```
toolbox.register("individualCreator", \
                 tools.initRepeat, \
                 creator.Individual, \
                 toolbox.zeroOrOne, ONE_MAX_LENGTH)
```

8. Lastly, we must register the populationCreator operator, which creates a list of individuals. This definition also uses the initRepeat operator, with the following arguments:

    - The list class as the container type

    - The individualCreator operator defined earlier as the function that's used to generate the objects in the list

    The last argument for initRepeat – the number of objects we want to generate – is not given here. This means that when using the populationCreator operator, this argument will be expected and used to determine the number of individuals that are created – in other words, the population size:

    ```
    toolbox.register("populationCreator", \
                     tools.initRepeat, \
                     list, toolbox.individualCreator)
    ```

9. To facilitate the fitness calculation (or evaluation, in DEAP terminology), we must define a standalone function that accepts an instance of the Individual class and returns the fitness for it. Here, we defined a function named oneMaxFitness that computes the number of 1s in the individual. Since the individual is essentially a list with values of either 1 or 0, the Python sum() function can be used for this purpose:

    ```
    def oneMaxFitness(individual):
        return sum(individual), # return a tuple
    ```

Tip

As mentioned previously, fitness values in DEAP are represented as tuples, and therefore a comma needs to follow when a single value is returned.

---

10. Next, we must define the evaluate operator as an alias to the oneMaxfitness() function we defined earlier. As you will see later, using the evaluate alias to calculate the fitness is a DEAP convention:

    ```
    toolbox.register("evaluate", oneMaxFitness)
    ```

    As we mentioned in the previous section, the genetic operators are typically created by aliasing existing functions from the tools module and setting the argument values as needed. Here, we chose the following:

    - Tournament selection with a tournament size of 3

    - Single-point crossover

    - Flip-bit mutation

Note the `indpb` parameter of the `mutFlipBit` function. This function iterates over all the attributes of the individual – a list containing values of 1s and 0s in our case – where each attribute will use this argument value as the probability of flipping (applying the `not` operator to) the attribute value. This value is independent of the mutation probability, which is set by the `P_MUTATION` constant that we defined earlier and has not been used yet. The mutation probability serves to decide whether the `mutFlipBit` function is called for a given individual in the population:

```
toolbox.register("select",tools.selTournament,\
                tournsize=3)
toolbox.register("mate", tools.cxOnePoint)
toolbox.register("mutate", tools.mutFlipBit,\
                indpb=1.0/ONE_MAX_LENGTH)
```

We are finally done with our settings and definitions. Now, we're ready to start the genetic flow, as described in the next section.

Evolving the solution

The genetic flow is implemented in the `main()` function, as described in the following steps:

1. We start the flow by creating the initial population using the `populationCreator` operator we defined earlier, and then using the `POPULATION_SIZE` constant as the argument for this operator. The `generationCounter` variable, which will be used later, is initialized here as well:

    ```
    population = toolbox.populationCreator(n=POPULATION_SIZE)
    generationCounter = 0
    ```

2. To calculate the fitness for each individual in the initial population, we can use the Python `map()` function to apply the `evaluate` operator to each item in the population. As the `evaluate` operator is an alias for the `oneMaxFitness()` function, the resulting `iterable` consists of the calculated fitness tuple of each individual. It is then converted into a `list` type of tuples:

    ```
    fitnessValues = list(map(toolbox.evaluate,\
                        population))
    ```

3. Since the items of `fitnessValues` match those in the population (which is a list of individuals), we can use the `zip()` function to combine them and assign the matching fitness tuple to each individual:

    ```
    for individual, fitnessValue in zip(population, fitnessValues):
        individual.fitness.values = fitnessValue
    ```

4. Next, since we have single-objective fitness, we must extract the first value out of each fitness to gather statistics:

```
fitnessValues = [
    individual.fitness.values[0] for individual in population
]
```

5. The statistics that are collected will be the max fitness value and the mean (average) fitness value for each generation. Two lists will be used for this purpose. Let's create them:

```
maxFitnessValues = []
meanFitnessValues = []
```

6. We are finally ready for the main loop of the genetic flow. At the top of the loop, we have the stopping conditions. As we decided earlier, one stopping condition will be set by putting a limit on the number of generations, and the other will be set by detecting that we have reached the best solution (a binary string containing all 1s):

```
while max(fitnessValues) < ONE_MAX_LENGTH and \
    generationCounter < MAX_GENERATIONS:
```

7. The generation counter is updated next. It is used by the stopping condition, as well as the `print` statements we will see soon:

```
generationCounter = generationCounter + 1
```

8. At the heart of the genetic algorithm are the *genetic operators*, which are applied next. The first is the *selection* operator, which can be applied using the `toolbox.select` operator we defined as the *tournament selection* earlier. Since we already set the tournament size when the operator was defined, we only need to send the population and its length as arguments now:

```
offspring = toolbox.select(population, len(population))
```

9. The selected individuals, now residing in a list called `offspring`, must be cloned so that we can apply the next genetic operators without affecting the original population:

```
offspring = list(map(toolbox.clone, offspring))
```

> **Important note**
> Despite the name `offspring`, these are still clones of individuals from the previous generation, and we still need to mate them using the `crossover` operator to create the actual offspring.

10. The next genetic operator is `crossover`. It was defined as the `toolbox.mate` operator earlier, and is aliasing a single-point crossover. We use Python extended slices to pair every even-indexed item of the `offspring` list with the one following it. Then, we utilize the

random() function to flip a coin using the *crossover probability* set by the P_CROSSOVER constant. This will decide if the pair of individuals will be crossed over or remain intact. Lastly, we delete the fitness values of the children since they have been modified and their existing fitness values are no longer valid:

```
for child1, child2 in zip(offspring[::2], offspring[1::2]):
    if random.random() < P_CROSSOVER:
        toolbox.mate(child1, child2)
        del child1.fitness.values
        del child2.fitness.values
```

> **Important note**
> The mate function takes two individuals as arguments and modifies them in place, meaning they don't need to be reassigned.

11. The final genetic operator to be applied is the *mutation*, which we registered earlier as the toolbox.mutate operator, and was set to be a *flip-bit* mutation operation. Iterating over all offspring items, the mutation operator will be applied at the probability set by the mutation probability constant, P_MUTATION. If the individual gets mutated, we must delete its fitness value (if it exists). This value could have carried over with the individual from the previous generation, and after mutation, it is no longer correct and needs to be recalculated:

```
for mutant in offspring:
    if random.random() < P_MUTATION:
        toolbox.mutate(mutant)
        del mutant.fitness.values
```

12. Individuals that were not crossed over or mutated remained intact, and therefore their existing fitness values, which were already calculated in a previous generation, don't need to be calculated again. The rest of the individuals will have this value empty. Now, we must find those fresh individuals using the Fitness class' valid property, then calculate the new fitness for them similarly to how to did the original calculation for fitness values:

```
freshIndividuals = [
    ind for ind in offspring if not ind.fitness.valid]
freshFitnessValues = list(map(toolbox.evaluate,
    freshIndividuals))

for individual, fitnessValue in zip(freshIndividuals,
    freshFitnessValues
):
    individual.fitness.values = fitnessValue
```

13. Now that the genetic operators are done, it is time to replace the old population with the new one:

```
population[:] = offspring
```

14. Before we continue to the next round, the current fitness values are collected to allow for statistical gathering. Since the fitness value is a (single element) tuple, we need to select the [0] index:

```
fitnessValues = [ind.fitness.values[0] for ind in population]
```

15. The max and mean fitness values are then found, at which point their values get appended to the statistics accumulators and a summary line is printed out:

```
maxFitness = max(fitnessValues)
meanFitness = sum(fitnessValues) / len(population)
maxFitnessValues.append(maxFitness)
meanFitnessValues.append(meanFitness)

print(f"- Generation {generationCounter}:
Max Fitness = {maxFitness}, \
    Avg Fitness = {meanFitness}")
```

16. In addition, we must locate the index of the (first) best individual using the max fitness value we just found and print this individual out:

```
best_index = fitnessValues.index(max(fitnessValues))
print("Best Individual = ", *population[best_index], "\n")
```

17. Once a stopping condition is activated and the genetic algorithm flow concludes, we can use the statistics accumulators to plot a couple of graphs using the matplotlib library. We can use the following code snippet to draw a graph illustrating the progress of the best and average fitness values throughout the generations:

```
plt.plot(maxFitnessValues, color='red')
plt.plot(meanFitnessValues, color='green')
plt.xlabel('Generation')
plt.ylabel('Max / Average Fitness')
plt.title('Max and Average fitness over Generations')
plt.show()
```

We are finally ready to test our first genetic algorithm – let's run it to find out if it finds the OneMax solution.

Running the program

When running the program described in the previous sections, we get the following output:

```
- Generation 1: Max Fitness = 65.0, Avg Fitness = 53.575
Best Individual = 1 1 0 1 0 1 0 0 1 0 0 0 1 1 1 0 1 0 0 1 0 1 0 0 0 1
1 1 1 1 0 1 1 1 1 0 1 0 1 1 1 1 0 0 1 1 11110111101111111000 0 1 0 1
0 1 1 1 0 1 1 0 0 0 1 1 1 0011111111111100
...
- Generation 40: Max Fitness = 100.0, Avg Fitness = 98.29
```

```
Best Individual = 1 1 1 1 1 1 1 1 1 1 1 1 1 1 1 1 1 1 1 1 1 1 1 1 1 1 1
1 1 1 1 1 1 1 1 1 1 1 1 1 1 1 1 1 1 1 1 1 11111111111111111111111 1 1 1 1
1 1 1 1 1 1 1 1 1 1 1 1 1 11111111111111111
```

As you can see, after 40 generations, an *all-1* solution was found, which yielded a fitness of 100 and stopped the genetic flow. The *average fitness*, which started at a value of around 53, ended at a value close to 100.

The graph that's plotted by `matplotlib` is shown here:

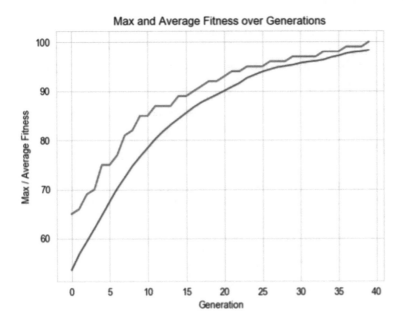

Figure 3.1: Stats of the program solving the OneMax problem

This plot illustrates how max fitness (the red line) increases over generations with incremental leaps, while the average fitness (the green line) keeps progressing smoothly.

Now that we've solved the OneMax problem using the DEAP framework, let's move on to the next section and find out how we can make our code more concise.

Using built-in algorithms

The DEAP framework comes with several built-in evolutionary algorithms provided by the `algorithms` module. One of them, `eaSimple`, implements the genetic algorithm flow we have been using, and can replace most of the code we had in the main method. Other useful DEAP objects, `Statistics` and `Logbook`, can be used for statistics gathering and printing, as we will soon see.

The program described in this section implements the same solution to the OneMax problem as the program discussed in the previous section but with less code. The only differences can be found in the `main` method. We will describe these differences in the following code snippets.

The complete program can be found here: `https://github.com/PacktPublishing/Hands-On-Genetic-Algorithms-with-Python-Second-Edition/blob/main/chapter_03/02_OneMax_short.py`.

The Statistics object

The first change we will make is in the way statistics are being gathered. To this end, we will now take advantage of the `tools.Statistics` class provided by DEAP. This utility enables us to create a `statistics` object using a key argument, which is a function that will be applied to the data on which the statistics are computed:

1. Since the data we plan to provide is the population of each generation, we'll set the key function to one that extracts the fitness value(s) from each individual:

```
stats = tools.Statistics(lambda ind: ind.fitness.values)
```

2. We can now register various functions that can be applied to these values at each step. In our example, we only use the max and mean NumPy functions, but others (such as min and std) can be registered as well:

```
stats.register("max", numpy.max)
stats.register("avg", numpy.mean)
```

As we will see soon, the collected statistics will be returned in an object called `logbook` at the end of the run.

The algorithm

Now, it's time for the actual flow. This is done with a single call to the `algorithms.eaSimple` method, one of the built-in evolutionary algorithms provided by the `algorithms` module of DEAP. When we call the method, we provide it with `population`, `toolbox`, and the `statistics` object, among other parameters:

```
population, logbook = algorithms.eaSimple(population, toolbox,
    cxpb=P_CROSSOVER,
    mutpb=P_MUTATION,
    ngen=MAX_GENERATIONS,
    stats=stats,
    verbose=True)
```

The `algorithms.eaSimple` method assumes that we previously used `toolbox` to register the following operators – `evaluate`, `select`, `mate`, and `mutate` – something we did when we created the original program. The stopping condition here is set by the value of `ngen`, which specifies the number of generations to run the algorithm for.

The logbook

When the flow is done, the algorithm returns two objects – the final population and a `logbook` object containing the statistics that were collected. We can now extract the desired statistics from the logbook using the `select()` method so that we can use them for plotting, as we did previously:

```
maxFitnessValues, meanFitnessValues = logbook.select("max", "avg")
```

We are now ready to run this slimmer version of the program.

Running the program

When running the program with the same parameter values and settings that we used previously, the printouts will be as follows:

```
gen    nevals   max    avg
0      200      61     49.695
1      193      65     53.575


...


39     192      99     98.04
40     173      100    98.29


...


49     187      100    99.83
50     184      100    99.89
```

These printouts are automatically generated by the `algorithms.eaSimple` method, according to the way we defined the `statistics` object sent to it, as the `verbose` argument was set to `True`.

The results are numerically similar to what we saw in the previous program, with two differences:

- There is a printout for generation 0; this was not included in the previous program.

- The genetic flow here continues to the 50th generation as this was the only stopping condition. In the previous program, there was an additional stopping condition that stopped the flow at the 40th generation since the best solution (that we happened to know beforehand) was reached.

We can observe the same behavior in the new graph plot. This graph is similar to the one we saw before but it continues to the 50th generation, even though the best result was already reached at the 40th generation:

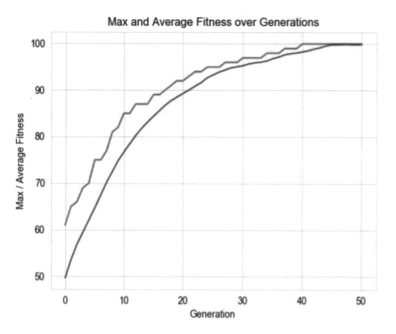

Figure 3.2: Stats of the program solving the OneMax problem using the built-in algorithm

Consequently, starting at the 40th generation, the value of the best fitness (the red line) no longer changes, while the average fitness (the green line) keeps climbing until it almost reaches the same max value. This means that by the end of this run, nearly all individuals are identical to the best one.

Adding the hall of fame feature

One additional feature of the built-in `algorithms.eaSimple` method is the **hall of fame** (or **HOF** for short). Implemented in the `tools` module, the `HallOfFame` class can be used to retain the best individuals that ever existed in the population during the evolution, even if they have been lost at some point due to selection, crossover, or mutation. The hall of fame is continuously sorted so that the first element is the first individual that has the best fitness value ever seen.

The complete program containing the code snippets shown in this section can be found here: https:// github.com/PacktPublishing/Hands-On-Genetic-Algorithms-with-Python-Second-Edition/blob/main/chapter_03/03_OneMax_short_hof.py.

To add the hall of fame functionality, let's make a few modifications to the previous program:

1. We start by defining a constant for the number of individuals we want to keep in the hall of fame. We will add this line to the constant definition section:

    ```
    HALL_OF_FAME_SIZE = 10
    ```

2. Just before calling the eaSimple algorithm, we'll create the HallOfFame object with that size:

    ```
    hof = tools.HallOfFame(HALL_OF_FAME_SIZE)
    ```

3. The HallOfFame object is sent as an argument to the eaSimple algorithm, which internally updates it during the run of the genetic algorithm flow:

    ```
    population, logbook = algorithms.eaSimple(\
        population, toolbox, cxpb=P_CROSSOVER, \
        mutpb=P_MUTATION, ngen=MAX_GENERATIONS, \
        stats=stats, halloffame=hof, verbose=True)
    ```

4. When the algorithm is done, we can use the HallOfFame object's items attribute to access the list of individuals who were inducted into the hall of fame:

    ```
    print("Hall of Fame Individuals = ", *hof.items, sep="\n")
    print("Best Ever Individual = ", hof.items[0])
    ```

    The printed results look as follows – the best individual consists of all 1s, followed by various individuals that have a 0 value in various locations:

    ```
    Hall of Fame Individuals =
    [1, 1, 1, 1, 1, 1, 1, 1, 1, 1, 1, 1, 1, 1, 1, 1, 1, 1, 1, 1,
    1, 1, 1, 1,1, 1, 1, 1, 1, 1, 1, 1, 1, 1, 1, 1, 1, 1, 1, 1, 1,
    1, 1, 1, 1, 1, 1, 1,1, 1, 1, 1, 1, 1, 1, 1, 1, 1, 1, 1, 1, 1,
    1, 1, 1, 1, 1, 1, 1, 1, 1, 1,1, 1, 1, 1, 1, 1, 1, 1, 1, 1, 1,
    1, 1, 1, 1, 1, 1, 1, 1, 1, 1, 1, 1, 1]  [1, 1, 1, 1, 1, 1, 1, 1,
    1, 1, 1, 1, 1, 1, 1, 0, 1, 1, 1, 1, 1, 1, 1, 1, 1, 1, 1, 1, 1,
    1, 1, 1, 1, 1, 1, 1, 1, 1, 1, 1, 1, 1, 1, 1, 1, 1, 1, 1, 1, 1,
    1, 1, 1, 1, 1, 1, 1, 1, 1, 1, 1, 1, 1, 1, 1, 1, 1, 1, 1, 1, 1,
    1, 1, 1, 1, 1, 1, 1, 1, 1, 1, 1, 1, 1, 1, 1, 1, 1, 1, 1, 1, 1,
    1, 1, 1, 1, 1, 1, 1, 1]

    ...
    ```

5. The best individual is the same one that was printed first previously:

    ```
    Best Ever Individual = [1, 1, 1, 1, ..., 0, ..., 1]
    ```

6. From now on, we will use these features – the statistics object and logbook, the built-in eaSimple algorithm, and HallOfFame – in all the programs we create.

Now that we've learned how to use the inbuilt algorithms, we'll experiment with them to find their differences and find the best algorithm for various uses.

Experimenting with the algorithm's settings

We can now experiment with the various settings and definitions we placed into the program and observe any changes in their behavior and results.

In each of the following subsections, we'll start from the original program settings and make one or more changes. You are encouraged to experiment with making your own modifications, as well as combining several modifications to be made to the same program.

Bear in mind that the effects of changes we make may be specific to the problem at hand – a simple *OneMax*, in our case – and may be different for other types of problems.

Population size and number of generations

We will start our experimentation by making modifications to the **population size** and the number of generations used by the genetic algorithm:

1. The size of the population is determined by the POPULATION_SIZE constant. We will start by increasing the value of this constant from 200 to 400:

    ```
    POPULATION_SIZE = 400
    ```

    This modification accelerates the genetic flow. The best solution is now found after 22 generations, as shown in the following figure:

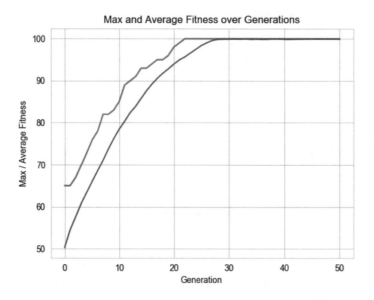

Figure 3.3: Stats of the program solving the OneMax problem after increasing the population size to 400

2. Next, we will try reducing the population size to 100:

```
POPULATION_SIZE = 100
```

3. This modification slows down the convergence of the algorithm, which will no longer reach the best possible value after 50 generations:

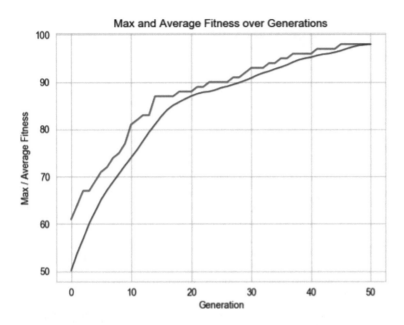

Figure 3.4: Stats of the program solving the OneMax problem after decreasing the population size to 100

4. To compensate, let's try increasing the value of MAX_GENERATIONS to 80:

```
MAX_GENERATIONS = 80
```

5.   We find that the best solution is now reached after 68 generations:

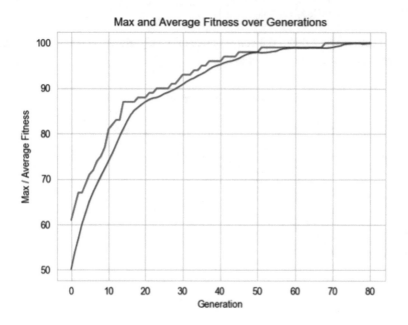

Figure 3.5: Stats of the program solving the OneMax problem
after increasing the number of generations to 80

This behavior is typical of genetic-algorithm-based solutions – increasing the population will require fewer generations to reach a solution. However, the computational and memory requirements increase with the population size, and we typically aspire to find a moderate population size that will provide a solution within a reasonable amount of time.

Crossover operator

Let's reset our changes and go back to the original settings (50 generations, population size 200). We are now ready to experiment with the **crossover** operator, which is responsible for creating offspring from parent individuals.

Changing the crossover type from a **single-point** to a **two-point** crossover is simple as we now define the `mate` operator as follows:

```
toolbox.register("mate", tools.cxTwoPoint)
```

The algorithm now finds the best solution after only 27 generations:

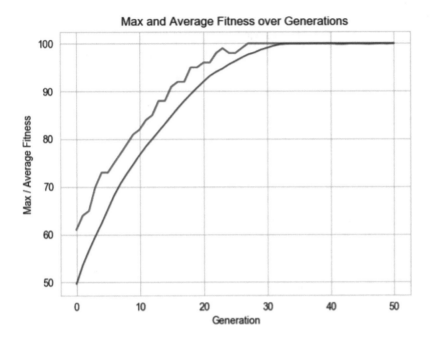

Figure 3.6: Stats of the program solving the OneMax problem after switching to a two-point crossover

This behavior is typical of genetic algorithms that utilize binary string representation as two-point crossover provides a more versatile way to combine two parents and mix their genes in comparison to the single-point crossover.

Mutation operator

We will now reset our changes again as we get ready to experiment with the **mutation** operator, which is responsible for introducing random modifications to offspring:

1. We will start by increasing the value of the P_MUTATION constant to 0.9. This results in the following plot:

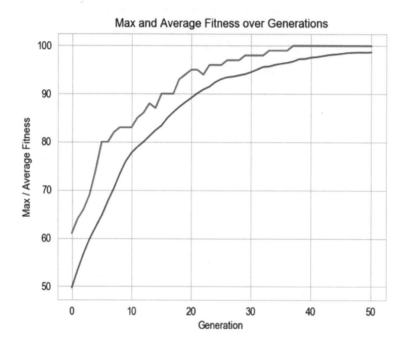

Figure 3.7: Stats of the program solving the OneMax problem
after increasing the mutation probability to 0.9

The results may seem surprising at first as increasing the mutation rate typically causes the algorithm to behave erratically, while here, the effect is seemingly unnoticeable. However, recall that there is another mutation-related parameter in our algorithm, indpb, which is an argument of the specific mutation operator we used here – mutFlipBit:

```
toolbox.register("mutate", tools.mutFlipBit, \
    indpb=1.0/ONE_MAX_LENGTH)
```

While the value of P_MUTATION determines the probability of an individual being mutated, indpb determines the probability of each bit in a given individual being flipped. In our program, we set the value of indpb to 1.0/ONE_MAX_LENGTH, which means that, on average, a single bit will be flipped in a mutated solution. For our 100-bit-long OneMax problem, this seems to limit the effect of the mutation, regardless of the P_MUTATION constant value.

2. Now, let's increase the value of indpb tenfold, as follows:

```
toolbox.register("mutate", tools.mutFlipBit, \
    indpb=10.0/ONE_MAX_LENGTH)
```

The result of running the algorithm with this value is somewhat erratic, as shown in the following figure:

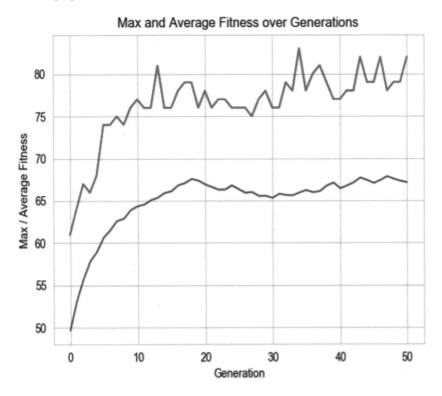

Figure 3.8: Stats of the program after a tenfold increase in the per-bit mutation probability

The figure indicates that while at first, the algorithm can improve the results, it quickly gets stuck in a state of oscillations without being able to make significant improvements.

3.   Increasing the `indpb` value further, to `50.0/ONE_MAX_LENGTH`, results in the following, unstable-looking, graph:

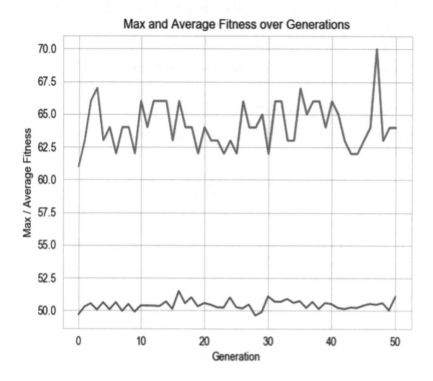

Figure 3.9: Stats of the program after a fifty-fold increase in the per-bit mutation probability

As evident from this plot, the genetic algorithm has turned into the equivalent of a random search – it may stumble upon the best solution by chance, but it doesn't make any progress toward better solutions.

Selection operator

Next, we'll look at the **selection** operator. First, we'll change the tournament size to see the combined effect of this parameter with the mutation probability. Then, we'll look at using *roulette* selection instead of *tournament* selection.

Tournament size and relation to mutation probability

Once again, we'll start by changing back to the original settings of the program before we make new modifications and run some experiments:

1. First, we'll modify the `tournamentSize` parameter of the tournament selection algorithm and change it to **2** (instead of the original value of **3**):

    ```
    toolbox.register("select", tools.selTournament, tournsize=2)
    ```

    This doesn't seem to have a noticeable effect on the algorithm's behavior:

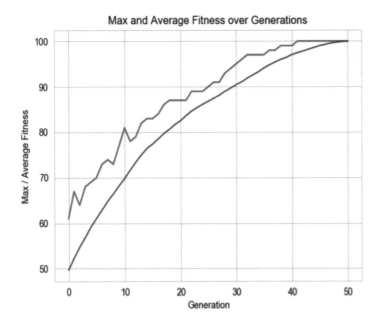

Figure 3.10: Stats of the program solving the OneMax problem after decreasing the tournament size to 2

2. What if we increase the tournament size to a very large value, say `100`?

3. Let's see:

```
toolbox.register("select", tools.selTournament, tournsize=100)
```

The algorithm still behaves well and finds the best solution in less than 40 generations. One noticeable effect is that the max fitness now closely resembles the average fitness, as shown in the following graph:

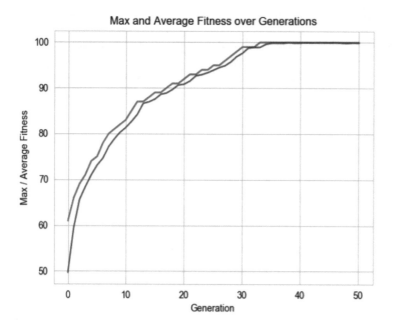

Figure 3.11: Stats of the program after increasing the tournament size to 100

This behavior occurs because when the tournament size increases, the chance of weak individuals being selected diminishes, and better solutions tend to take over the population. In real-life problems, this takeover might cause suboptimal solutions to saturate the population and prevent the best solution from being found (a phenomenon known as **premature convergence**). However, in the case of the simple OneMax problem, this doesn't seem to be an issue. A possible explanation is that the mutation operator provides enough diversity to keep the solutions moving in the right direction.

4. To put this explanation to the test, let's reduce the mutation probability tenfold, to `0.01`:

    ```
    P_MUTATION = 0.01
    ```

    If we run the algorithm again, we'll see that the results stop improving soon after the start of the algorithms, and then improve at a much slower pace, with an occasional improvement here and there. The overall results are far worse than the previous run as the best fitness is around 80 rather than 100:

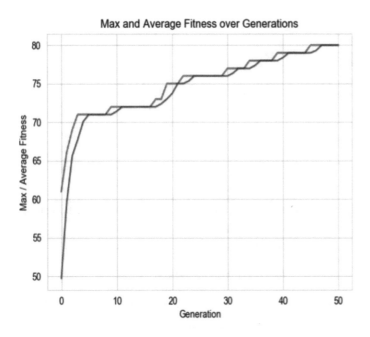

Figure 3.12: Stats of the program with a tournament size of 100 and a mutation probability of 0.01

This interpretation is that due to the large tournament size, the best individuals from the initial population take over within a small number of generations, which shows in the initial quick increase of both graphs in the plot. After that, only an occasional mutation in the right direction – one that flips a 0 to 1 –creates a better individual; this is indicated in the plot by a jump of the red line. Soon after, this individual takes over the entire population again, where the green line catches up with the red one.

5. To make this situation even more extreme, we can further reduce the mutation rate:

    ```
    P_MUTATION = 0.001
    ```

We can now see the same general behavior, but since mutations are very rare, the improvements are few and far between:

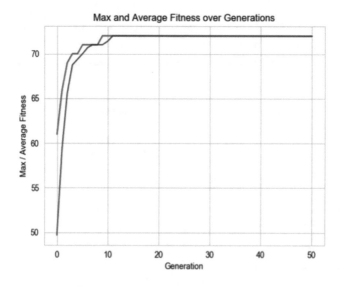

Figure 3.13: Stats of the program with a tournament size of 100 and a mutation probability of 0.001

6.   Now, if we increase the number of generations to 500, we can see this behavior more clearly:

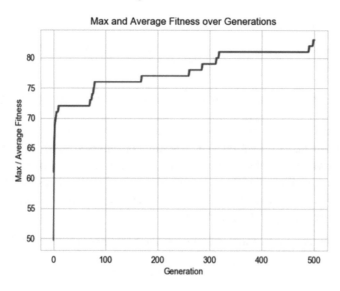

Figure 3.14: Stats of the program with a tournament size of 100 and
a mutation probability of 0.001, over 500 generations

7. Just out of curiosity, let's dial back the tournament size to 3 again and restore the number of generations to 50, leaving the small mutation rate in place:

```
MAX_GENERATIONS = 50
toolbox.register("select", tools.selTournament, tournsize=3)
```

The resulting plot is a lot closer to the original one:

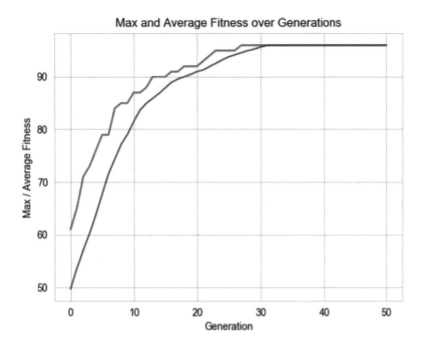

Figure 3.15: Stats of the program with a tournament size of 3 and a
mutation probability of 0.001, over 50 generations

Here, it seems that a takeover occurred as well, but far later, around generation 30, when the best fitness was already close to the maximum value of 100. Here, a more reasonable mutation rate would help us find the best solution, as happened with the original settings.

Roulette wheel selection

Let's go back to the original settings once more, in preparation for our last experiment, as we will now try replacing the tournament selection algorithm with **roulette wheel selection**, which was described in *Chapter 2, Understanding the Key Components of Genetic Algorithms*. This is done as follows:

```
toolbox.register("select", tools.selRoulette)
```

This change seems to harm the algorithm's results. As the following plot shows, there are numerous points in time where the best solution is forgotten as a result of the selection, and the max fitness value decreases, at least temporarily, although the average fitness value keeps increasing. This is because the roulette selection algorithm selects individuals with a probability proportionate to their fitness; when the differences between the individuals are relatively small, there is a better chance for weaker individuals to be selected, in comparison to the tournament selection we had before:

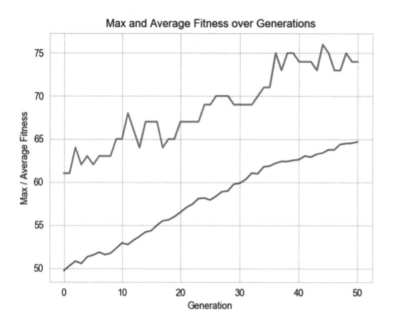

Figure 3.16: Stats of the program when using roulette wheel selection

To compensate for this behavior, we can use the **elitist approach** mentioned in *Chapter 2, Understanding the Key Components of Genetic Algorithms*. This approach allows a certain number of the best individuals from the current generation to carry over to the next generation unaltered and prevents them from being lost. In the next chapter, we will explore applying the elitist approach when using the DEAP library.

Summary

In this chapter, you were introduced to **DEAP** – a versatile evolutionary computation framework that will be used in the rest of this book to solve real-life problems using genetic algorithms. You learned about DEAP's `creator` and `toolbox` modules, and how to use them to create the various components needed for the genetic algorithm's flow. DEAP was then used to write two versions of a Python program that solves the *OneMax* problem, the first with full implementation of the genetic algorithm flow, and the other – more concise – taking advantage of the built-in algorithms of the framework. A third version of the program introduced the HOF feature offered by DEAP. We then experimented with various settings of the genetic algorithm and discovered the effects of changing the population size, as well as modifying the *selection*, *crossover*, and *mutation* operators.

In the next chapter, expanding on what we learned in this chapter, we will start solving real-life combinatorial problems, including the *traveling salesman problem* and the *vehicle routing problem*, using DEAP-based Python programs.

Further reading

For more information, please refer to the following resources:

- DEAP documentation: `https://deap.readthedocs.io/en/master/`
- DEAP source code on GitHub: `https://github.com/DEAP/deap`

4

Combinatorial Optimization

In this chapter, you will learn how genetic algorithms can be utilized in combinatorial optimization applications. We will start by describing search problems and combinatorial optimization, and outline several hands-on examples of combinatorial optimization problems. We will then analyze each of these problems and match them with Python-based solutions using the DEAP framework. The optimization problems we'll cover are the well-known knapsack problem, the **traveling salesman problem** (**TSP**), and the **vehicle routing problem** (**VRP**). As a bonus, we will cover the topics of genotype-to-phenotype mapping and exploration versus exploitation.

By the end of this chapter, you will be able to do the following:

- Understand the nature of search problems and combinatorial optimization
- Solve the knapsack problem using a genetic algorithm coded with the DEAP framework
- Solve the TSP using a genetic algorithm coded with the DEAP framework
- Solve the VRP using a genetic algorithm coded with the DEAP framework
- Understand genotype-to-phenotype mapping
- Gain familiarity with the concept of exploration versus exploitation and its relation to elitism

Technical requirements

In this chapter, we will be using Python 3 with the following supporting libraries:

- `deap`
- `numpy`
- `matplotlib`
- `seaborn`

> **Important note**
>
> If you use the `requirements.txt` file provided (see Chapter 3), these libraries will already be in your environment.

In addition, we will be using the benchmark data from the *Rosetta Code* (`https://rosettacode.org/wiki/Rosetta_Code`) and *TSPLIB* (`http://comopt.ifi.uni-heidelberg.de/software/TSPLIB95/`) web pages.

The programs that will be used in this chapter can be found in this book's GitHub repository: `https://github.com/PacktPublishing/Hands-On-Genetic-Algorithms-with-Python-Second-Edition/tree/main/chapter_04`. Check out the following video to see the Code in Action: `https://packt.link/OEBOd`

Search problems and combinatorial optimization

One common area of applying genetic algorithms is *search problems*, which have important applications in fields such as logistics, operations, artificial intelligence, and machine learning. Examples include determining the optimal routes for package delivery, designing hub-based airline networks, managing investment portfolios, and assigning passengers to available drivers in a fleet of taxis.

Search algorithms focus on solving a problem through methodic evaluation of **states** and **state transitions**, aiming to find a path from the initial state to a desirable final (or "goal") state. Typically, there is a **cost** or a **gain** involved in every state transition, and the objective of the corresponding search algorithm is to find a path that minimizes the cost or maximizes the gain. Since the optimal path is one of many possible ones, this kind of search is related to *combinatorial optimization*, a topic that involves finding an optimal object from a finite, yet often extremely large, set of possible objects.

These concepts will be illustrated as we get acquainted with the *knapsack problem*, which is the main focus of the next section.

Solving the knapsack problem

Think of the familiar situation of packing for a long trip. There are many items that you would like to take with you, but you are limited by the capacity of your suitcase. In your mind, each item has a certain value it will add to your trip; at the same time, it has a size (and weight) associated with it, and it will compete with other items over the available space in your suitcase. This situation is just one of many real-life examples of the *knapsack problem*, which is considered one of the oldest and most investigated combinatorial search problems.

More formally, the knapsack problem consists of the following components:

- A set of **items**, each of them associated a certain **value** and a certain **weight**
- A **bag/sack/container** (the "knapsack") of a certain **weight capacity**

Our goal is to come up with a group of selected items that will provide the maximum total value, without exceeding the total weight capacity of the bag.

In the context of search algorithms, each subset of the items represents a state, and the set of all possible item subsets is considered the state space. For an instance of the knapsack 0-1 problem with n items, the size of the state space is 2 n, which can quickly grow very large, even for a modest value of n.

In this (original) version of the problem, each item can only be included once or not at all, and therefore it is sometimes referred to as the **knapsack 0-1** problem. However, it can be expanded into other variants – for example, where items can be included multiple times (limited or unlimited) or where multiple knapsacks with varying capacities are present.

Applications of knapsack problems appear in many real-world processes that involve resource allocation and decision-making, such as selecting investments when building an investment portfolio, minimizing the waste when cutting raw materials, and getting the "most bang for your buck" when selecting which questions to answer in a timed test.

To get our hands dirty with a knapsack problem, we will look at a widely known example.

The Rosetta Code knapsack 0-1 problem

The *Rosetta Code* website (`rosettacode.org`) provides a collection of programming tasks, each with solutions in numerous languages. One of these tasks, described at `rosettacode.org/wiki/Knapsack_problem/0-1`, is a knapsack 0-1 problem where a tourist needs to decide which items to pack for their weekend trip. The tourist has 22 items they can choose from; each item is assigned by the tourist with some value that represents its relative importance for the upcoming journey.

The weight capacity of the tourist's bag in this problem is **400**. The list of items, along with their associated values and weights, is provided in the following table:

Item	Weight	Value
map	9	150
compass	13	35
water	153	200
sandwich	50	160
glucose	15	60
tin	68	45
banana	27	60
apple	39	40
cheese	23	30
beer	52	10

Item	Weight	Value
suntan cream	11	70
camera	32	30
T-shirt	24	15
trousers	48	10
umbrella	73	40
waterproof trousers	42	70
waterproof overclothes	43	75
note-case	22	80
sunglasses	7	20
towel	18	12
socks	4	50
book	30	10

Table 4.1: A list of Rosetta Code knapsack 0-1 items

Before we start solving this problem, we need to discuss one important matter – what's a potential solution?

Solution representation

When solving the knapsack 0-1 problem, a straightforward way to represent a solution is using a list of binary values. Every entry in that list corresponds to one of the items in the problem. For the Rosetta Code problem, a solution can be represented using a list of 22 integers of the values 0 or 1. A value of 1 represents picking the corresponding item, while a value of 0 means that the item hasn't been picked. When applying the genetic algorithms approach, this list of binary values is going to be used as the chromosome.

However, we have to remember that the total weight of the chosen items cannot exceed the capacity of the knapsack. One way to incorporate this restriction into the solution is to wait until it gets evaluated. We then evaluate by adding the weights of the chosen items one by one, while ignoring any chosen item that will cause the accumulated weight to exceed the maximum allowed value. From the genetic algorithm's point of view, this means that the chromosome representation of an individual (*genotype*) may not entirely express itself when it gets translated into the actual solution (*phenotype*) as some of the 1 values in the chromosome may be ignored. This situation is sometimes referred to as **genotype-to-phenotype mapping**.

The solution representation we just discussed is implemented in the Python class described in the next subsection.

Python problem representation

To encapsulate the Rosetta Code knapsack 0-1 problem, we created a Python class called `Knapsack01Problem`. This class is contained in the `knapsack.py` file, which can be found at `https://github.com/PacktPublishing/Hands-On-Genetic-Algorithms-with-Python-Second-Edition/blob/main/chapter_04/knapsack.py`.

The class provides the following methods:

- `__init_data()`: This initializes the `RosettaCode.org` knapsack 0-1 problem data by creating a list of tuples. Each tuple contains the name of an item, followed by its weight and its value.

- `getValue(zeroOneList)`: This calculates the value of the chosen items in the list while ignoring items that will cause the accumulating weight to exceed the maximum weight.

- `printItems(zeroOneList)`: This prints the chosen items in the list while ignoring items that will cause the accumulating weight to exceed the maximum weight.

The `main()` method of the class creates an instance of the `Knapsack01Problem` class. It then creates a random solution and prints out its relevant information. If we run this class as a standalone Python program, a sample output may look as follows:

```
Random Solution =
[1 1 1 1 0 0 0 0 1 1 1 0 1 0 0 0 1 0 0 0 0]

- Adding map: weight = 9, value = 150, accumulated weight = 9,
accumulated value = 150

- Adding compass: weight = 13, value = 35, accumulated weight = 22,
accumulated value = 185

- Adding water: weight = 153, value = 200, accumulated weight = 175,
accumulated value = 385

- Adding sandwich: weight = 50, value = 160, accumulated weight = 225,
accumulated value = 545

- Adding glucose: weight = 15, value = 60, accumulated weight = 240,
accumulated value = 605

- Adding beer: weight = 52, value = 10, accumulated weight = 292,
accumulated value = 615
```

```
- Adding suntan cream: weight = 11, value = 70, accumulated weight =
303, accumulated value = 685

- Adding camera: weight = 32, value = 30, accumulated weight = 335,
accumulated value = 715

- Adding trousers: weight = 48, value = 10, accumulated weight = 383,
accumulated value = 725

- Total weight = 383, Total value = 725
```

Note that the last occurrence of 1 in the random solution, representing the note-case item, fell victim to the *genotype-to-phenotype mapping* discussed in the previous subsection. As this item's weight is 22, it would cause the total weight to exceed 400. As a result, this item was not included in the solution.

This random solution, as you may expect, is far from being optimal. Let's try and find the optimal solution for this problem using a genetic algorithm.

Genetic algorithm solution

To solve our knapsack 0-1 problem using a genetic algorithm, we created the 01-solve-knapsack. py Python program located at https://github.com/PacktPublishing/Hands-On-Genetic-Algorithms-with-Python-Second-Edition/blob/main/chapter_04/01_solve_knapsack.py.

As a reminder, the chromosome representation we decided to use here is a list of integers with the values of 0 or 1. This makes our problem, from the point of view of the genetic algorithm, similar to the OneMax problem we solved in the previous chapter. The genetic algorithm doesn't care what the chromosome represents (also known as the *phenotype*) – a list of items to pack, a sequence of Boolean equation coefficients, or perhaps an actual binary number; it is only concerned with the chromosome itself (the *genotype*) and the fitness value of that chromosome. Mapping the chromosome to the solution it represents is carried out by the fitness evaluation function, which is implemented outside the genetic algorithm. In our case, this chromosome mapping and fitness calculation is implemented by the getValue() method, which is encapsulated within the Knapsack01Problem class.

The outcome of all this is that we can use the same genetic algorithm implementation that we used for the On-Max problem, with a few adaptations.

The following steps describe the main points of our solution:

1. First, we need to create an instance of the knapsack problem we would like to solve:

    ```
    knapsack = knapsack.Knapsack01Problem()
    ```

2. Then, we must instruct the genetic algorithm to use the `getValue()` method of that instance for fitness evaluation:

```
def knapsackValue(individual):
    return knapsack.getValue(individual),

toolbox.register("evaluate", knapsackValue)
```

3. The genetic operators that are used are compatible with the binary-list chromosome:

```
toolbox.register("select", tools.selTournament, tournsize=3)
toolbox.register("mate", tools.cxTwoPoint)
toolbox.register("mutate", tools.mutFlipBit,
    indpb=1.0/len(knapsack))
```

4. Once the genetic algorithm stops, we can use the `printItems()` method to pretty-print the best solution that was found:

```
best = hof.items[0]
print("-- Knapsack Items = ")
knapsack.printItems(best)
```

5. We can also tweak some of the parameters of the genetic algorithm. As this particular problem uses a binary string of length 22, it seems easier than the 100-length OneMax problem we previously solved, so we can probably reduce the population size and maximum number of generations.

    Upon running the algorithm for 50 generations, with a population size of 50, we get the following outcome:

```
-- Best Ever Individual = [1, 1, 1, 1, 1, 0, 1, 0, 0, 0, 1, 0,
0, 0, 0, 1, 1, 1, 1, 0, 1, 1]


-- Best Ever Fitness = 1030.0


-- Knapsack Items =

- Adding map: weight = 9, value = 150, accumulated weight = 9,
accumulated value = 150


- Adding compass: weight = 13, value = 35, accumulated weight =
22, accumulated value = 185


- Adding water: weight = 153, value = 200, accumulated weight =
175, accumulated value = 385
```

```
- Adding sandwich: weight = 50, value = 160, accumulated weight
= 225, accumulated value = 545

- Adding glucose: weight = 15, value = 60, accumulated weight =
240, accumulated value = 605

- Adding banana: weight = 27, value = 60, accumulated weight =
267, accumulated value = 665

- Adding suntan cream: weight = 11, value = 70, accumulated
weight = 278, accumulated value = 735

- Adding waterproof trousers: weight = 42, value = 70,
accumulated weight = 320, accumulated value = 805

- Adding waterproof overclothes: weight = 43, value = 75,
accumulated weight = 363, accumulated value = 880

- Adding note-case: weight = 22, value = 80, accumulated weight
= 385, accumulated value = 960

- Adding sunglasses: weight = 7, value = 20, accumulated weight
= 392, accumulated value = 980

- Adding socks: weight = 4, value = 50, accumulated weight =
396, accumulated value = 1030

- Total weight = 396, Total value = 1030
```

The total value of 1030 is the known optimal solution for this problem.

Here, too, we can see that the last occurrence of 1 in the chromosome of the best individual, representing the book item, was sacrificed to the actual solution in the mapping to keep the accumulated weight from exceeding the limit of 400.

The following figure, which depicts the max and average fitness over the generations, indicates that the best solution was found in less than 10 generations:

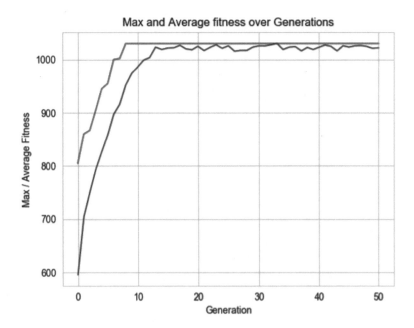

Figure 4.1: Stats of the program solving the knapsack 0-1 problem

In the next section, we will shift gears and tackle a more involved, yet still classic, combinatorial search task known as the TSP.

Solving the TSP

Imagine that you manage a small fulfillment center and need to deliver packages to a list of customers using a single vehicle. What's the best route for the vehicle to take so that you can visit all your customers and then return to the starting point? This is an example of the classic **TSP**.

The TSP dates back to 1930, and since then has been one of the most thoroughly studied problems in optimization. It is often used to benchmark optimization algorithms. The problem has many variants, but it was originally formulated after a traveling salesman who needs to take a trip that covers several cities:

> *"Given a list of cities and the distances between each pair of the cities, find the shortest possible path that goes through all the cities and then returns to the starting city."*

Using combinatorics, you could find that when given n cities, the number of possible paths that go through all cities is $(n - 1)!/2$.

The following figure shows the shortest path for the traveling salesperson problem that covers the 15 largest cities in Germany:

Figure 4.2: The shortest TSP path for the 15 largest cities in Germany.

Source: https://commons.wikimedia.org/wiki/File:TSP_Deutschland_3.png.

Image by Kapitän Nemo.

In this case, *n=15*, so the number of possible routes is *14!/2*, which amounts to the staggering number of 43,589,145,600.

In the context of search algorithms, each path (or partial path) through the cities represents a state, and the set of all possible paths is considered the state space. Each of the paths has a corresponding "cost" – the length (distance) of the path – and we are looking for the path that will minimize this distance.

As we pointed out, the state space is very large, even for a moderate number of cities, which can make it prohibitively expensive to evaluate every possible path. As a result, even though it is relatively easy to find a path that goes through all the cities, finding the *optimal* path can be very hard.

TSPLIB benchmark files

The **TSPLIB** is a library containing sample problems for the TSP based on the actual geographic locations of cities. The library is maintained by Heidelberg University, and relevant examples can be found here: http://comopt.ifi.uni-heidelberg.de/software/TSPLIB95/tsp/.

Two types of files can be found on this web page: files with the `.tsp.gz` suffix, each of them containing the description of a particular TSP problem, and the corresponding `.opt.tour.gz` files, containing the optimal solution for each problem.

The problem description files are text-based and white-space delimited. A typical file contains several informational lines, followed by city data. We are interested in files that include the x, y coordinates of the participating cities so that we can plot the cities and visualize their locations. For example, the contents of the `burma14.tsp.gz` file, once unzipped, look as follows (with some of the lines omitted here for brevity):

```
NAME: burma14
TYPE: TSP
...
NODE_COORD_SECTION
    1   16.47        96.10
    2   16.47        94.44
    3   20.09        92.54

    ...
   12   21.52        95.59
   13   19.41        97.13
   14   20.09        94.55
EOF
```

The interesting section for us is the lines between `NODE_COORD_SECTION` and `EOF`. In some of the files, `DISPLAY_DATA_SECTION` is used instead of `NODE_COORD_SECTION`.

Are we ready to solve a sample problem? Well, before we start doing that, we still need to figure out how a potential solution will be represented. This will be addressed in the next subsection.

Solution representation

When solving the TSP, the cities are typically represented by numbers from 0 to n-1, and possible solutions will be sequences of these numbers. A problem with five cities, for example, can have solutions of the form [0,1, 2, 3, 4], [2, 4, 3, 1, 0], and so on. Each solution can be evaluated by calculating and totaling the distances between each two subsequent cities, then adding the distance between the last city to the first one. Consequently, when applying the genetic algorithms approach to this problem, we can use a similar list of integers to serve as the chromosome.

The Python class described in the next subsection reads the contents of TSPLIB files and calculates the distances between each two cities. In addition, it calculates the total distance covered by a given potential solution using the list representation we just discussed.

Python problem representation

To encapsulate the TSP problem, we've created a Python class called `TravelingSalesmanProblem`. This class is contained in the `tsp.py` file and can be found at `https://github.com/PacktPublishing/Hands-On-Genetic-Algorithms-with-Python-Second-Edition/blob/main/chapter_04/tsp.py`.

The class provides the following private methods:

- `__create_data()`: This reads the desired TSPLIB file, extracts the coordinates of all cities, calculates the distances between every two cities, and uses them to populate a distance matrix (two-dimensional array). It then serializes the city locations and the calculated distances to disk using the `pickle` utility.

- `__read_data()`: This reads the serialized data and, if not available, calls `__create_data()` to prepare it.

These methods are invoked internally by the constructor, so the data is initialized as soon as the instance is created.

In addition, the class provides the following public methods:

- `getTotalDistance(indices)`: This calculates the total distance of the path described by the given list of city indices

- `plotData(indices)`: This clots the path described by the given list of city indices

The main method of the class exercises the class methods mentioned previously: first, it creates the `bayg29` problem (29 cities in Bavaria), then calculates the distance for the hard-coded optimal solution (as described in the matching `.opt.tour` file), and finally plots it. So, if we run this class as a standalone Python program, the output will be as follows:

```
Problem name: bayg29

Optimal solution = [0, 27, 5, 11, 8, 25, 2, 28, 4, 20, 1, 19, 9, 3,
14, 17, 13, 16, 21, 10, 18, 24, 6, 22, 7, 26, 15, 12, 23]

Optimal distance = 9074.147
```

The plot of the optimal solution looks as follows:

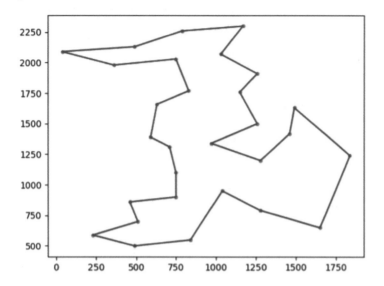

Figure 4.3: A plot of the optimal solution for the "bayg29" TSP. The red dots represent cities

Next, we will try to reach this optimal solution using a genetic algorithm.

Genetic algorithm solution

For our first attempt at solving the TSP using a genetic algorithm, we created the `02-solve-tsp-first-attempt.py` Python program, which is located at `https://github.com/PacktPublishing/Hands-On-Genetic-Algorithms-with-Python-Second-Edition/blob/main/chapter_04/02_solve_tsp_first_attempt.py`.

The main parts of our solution are described in the following steps:

1. The program starts by creating an instance of the `bayg29` problem, as follows:

   ```
   TSP_NAME = "bayg29"
   tsp = tsp.TravelingSalesmanProblem(TSP_NAME)
   ```

2. Next, we need to define the fitness strategy. Here, we want to minimize the distance, which translates to a single-objective minimizing `fitness` class that's defined using a single negative weight:

   ```
   creator.create("FitnessMin", base.Fitness, weights=(-1.0,))
   ```

3. As we discussed a little earlier, our choice of chromosome for the genetic algorithm is a list of integers from 0 to *n-1*, where *n* is the number of cities, representing the city indices. As an example, the optimal solution we saw earlier for the "bayg29" problem was represented with the following chromosome:

    ```
    [0, 27, 5, 11, 8, 25, 2, 28, 4, 20, 1, 19, 9, 3, 14, 17, 13,
    16, 21, 10, 18, 24, 6, 22, 7, 26, 15, 12, 23]
    ```

    The following code snippet is responsible for implementing this chromosome. It's explained further after:

    ```
    creator.create("Individual", array.array, typecode='i',
        fitness=creator.FitnessMin)

    toolbox.register("randomOrder", random.sample,
        range(len(tsp)), len(tsp))

    toolbox.register("individualCreator", tools.initIterate,
        creator.Individual, toolbox.randomOrder)

    toolbox.register("populationCreator", tools.initRepeat, list,
        toolbox.individualCreator)
    ```

    The `Individual` class is created first, extending an array of integers and augmenting it with the `FitnessMin` class.

    The `randomOrder` operator is then registered to provide the results of `random.sample()` invocation over a range defined by the length of the TSP problem (the number of cities, or *n*). This will result in a randomly generated list of indices between 0 and *n-1*.

    The `IndividualCreator` operator is created next. When called, it will invoke the `randomOrder` operator and iterate over the results to create a valid chromosome consisting of the city indices.

    The last operator, `populationCreator`, is created to produce a list of individuals using the `IndividualCreator` operator.

4. Now that the chromosome has been implemented, it's time to define the fitness evaluation function. This is carried out by the `tspDistance()` function, which directly utilizes the `getTotalDistance()` method of the `TravelingSalesmanProblem` class:

    ```python
    def tpsDistance(individual):
        return tsp.getTotalDistance(individual),   # return a tuple

    toolbox.register("evaluate", tpsDistance)
    ```

5. Next, we need to define the genetic operators. For the selection operator, we can use tournament selection with a tournament size of 3, as we did in previous cases:

```
toolbox.register("select", tools.selTournament, tournsize=3)
```

6. However, before picking the crossover and mutation operators, we need to remember that the chromosome we use is not just a list of integers but a list of indices (or an **ordered list**) that represent the order of the cities, and therefore we cannot just mix parts of two lists or arbitrarily change an index in the list. Instead, we need to use specialized operators that were designed to produce valid lists of indices. In *Chapter 2, Understanding the Key Components of Genetic Algorithms*, we examined several of these operators, including **ordered crossover** and **scramble mutation**. Here, we're using DEAP's corresponding implementations of these operators, cxOrdered and mutShuffleIndexes:

```
toolbox.register("mate", tools.cxOrdered)
toolbox.register("mutate", tools.mutShuffleIndexes,
    indpb=1.0/len(tsp))
```

7. Finally, it's time to invoke the genetic algorithm flow. Here, we use the default DEAP built-in eaSimple algorithm, with our default stats and halloffame objects to provide information we can display later:

```
population, logbook = algorithms.eaSimple(population, \
    toolbox, cxpb=P_CROSSOVER, mutpb=P_MUTATION, \
    ngen=MAX_GENERATIONS, stats=stats, halloffame=hof, \
    verbose=True)
```

Running this program with the constant values appearing at the top of the file (a population size of 300, 200 generations, a crossover probability of 0.9, and a mutation probability of 0.1) yields the following results:

```
-- Best Ever Individual = Individual('i', [0, 27, 11, 5, 20, 4, 8, 25,
2, 28, 1, 19, 9, 3, 14, 17, 13, 16, 21, 10, 18, 12, 23, 7, 26, 22, 6,
24, 15])

-- Best Ever Fitness = 9549.9853515625
```

The best fitness found (9549.98) is not too far from the known optimal distance of 9074.14.

The program then produces two plots. The first plot illustrates the path of the best individual found during the run:

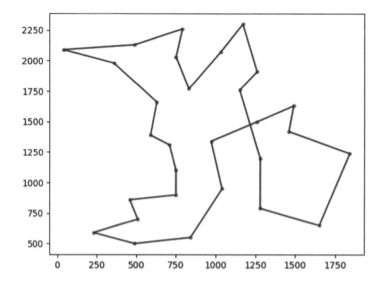

Figure 4.4: A plot of the best solution found by the first program attempting to solve the "bayg29" TSP

The second plot shows the statistics of the genetic flow. Note that this time, we chose to collect data for the *minimum* fitness value rather than the maximum as the objective of this problem is to minimize the distance:

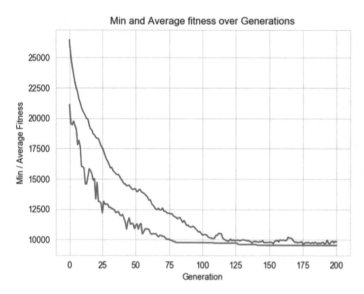

Figure 4.5: Stats of the first program attempting to solve the "bayg29" TSP

Now that we've found a good solution but not the best-known one, we can try and figure out ways to improve the results. For example, we can experiment with changing the population size, number of generations, and probabilities. We can also replace the genetic operators with other compatible ones. We can even change the random seed we set just to see the effect on the results or make multiple runs with different seeds. In the next section, we will try to use **elitism** combined with **enhanced exploration** to improve our results.

Improving the results with enhanced exploration and elitism

If we try to increase the number of generations in the previous program, we will realize that the solution doesn't improve – it's stuck in the (somewhat) suboptimal solution that was reached sometime before generation 200. This is shown in the following plot, which displays 500 generations:

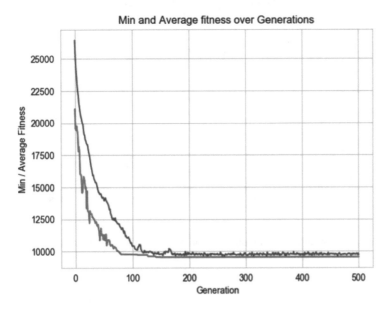

Figure 4.6: Stats of the first program, running for 500 generations

From that point on, the similarity between the average value and the best value indicates that this solution took over the population and therefore we will not see any improvement unless a lucky mutation turns up. In genetic algorithms terms, this means that **exploitation** has overpowered **exploration**. Exploitation generally means taking advantage of the current available results, while exploration emphasizes the search for new solutions. Striking a delicate balance between the two can lead to better results.

One way to increase exploration could involve reducing the tournament size of the tournament selection that's used from 3 to 2:

```
toolbox.register("select", tools.selTournament, tournsize=2)
```

As we discussed in *Chapter 2, Understanding the Key Components of Genetic Algorithms*, this will increase the chances of less successful individuals being selected. These individuals may carry the key to better future solutions. However, if we run the same program after making this change, the results are far from impressive – the best fitness value is over 13,000, and the best solution plot looks as follows:

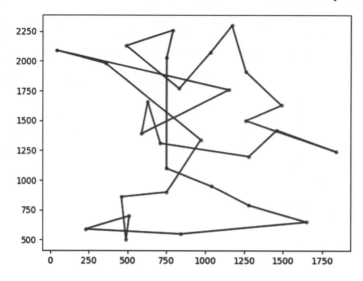

Figure 4.7: A plot of the best solution found by the program with the tournament size reduced to 2

These poor results can be explained using the statistics plot:

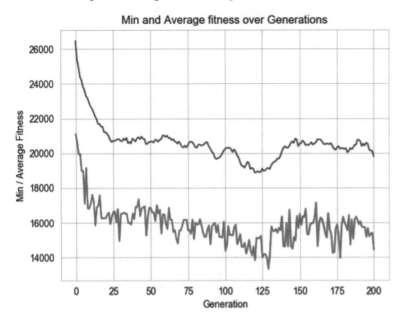

Figure 4.8: Stats of the program with the tournament size reduced to 2

This plot illustrates that we cannot retain the best solutions. As evident from the "noisy" graph, which keeps jumping between better values to worse values, good solutions tend to quickly get "lost" due to the more permissive selection scheme, which often enables lesser solutions to be selected. This means that we let exploration go too far, and to balance it out, we need to re-introduce a measure of exploitation into the mix. This can be done using the **elitism** mechanism, which was first introduced in *Chapter 2*.

Elitism enables us to keep the best solutions intact by letting them "skip" the genetic operators of selection, crossover, and mutation during the genetic flow. To implement elitism, we will have to go "under the hood" and modify DEAP's `algorithms.eaSimple()` algorithm as the framework doesn't provide a direct way to skip all three operators.

The modified algorithm, called `eaSimpleWithElitism()`, can be found in the `elitism.py` file, located at `https://github.com/PacktPublishing/Hands-On-Genetic-Algorithms-with-Python-Second-Edition/blob/main/chapter_04/elitism.py`.

The `eaSimpleWithElitism()` method is similar to the original `eaSimple()`, with the modification that the `halloffame` object is now used to implement an elitism mechanism. The individuals contained in the `halloffame` object are directly injected into the next generation and are not subject to the genetic operators of selection, crossover, and mutation. This is essentially the outcome of the following modifications:

- Instead of selecting several individuals equal to the population size, this number of selected individuals is reduced by the number of hall-of-fame individuals:

```
offspring = toolbox.select(population,
    len(population) - hof_size)
```

- After the genetic operators have been applied, the hall-of-fame individuals are added back into the population:

```
offspring.extend(halloffame.items)
```

We can now replace the call to `algorithms.eaSimple()` with a call to `elitism.eaSimpleWithElitism()`, without changing any of the parameters. Then, we'll set the HALL_OF_FAME_SIZE constant to 30, which means that we will always keep the best 30 individuals in the population.

The modified Python program, `03-solve-tsp.py` can be found at https://github.com/PacktPublishing/Hands-On-Genetic-Algorithms-with-Python-Second-Edition/blob/main/chapter_04/03_solve_tsp.py.

Upon running this new program, we are now able to hit the optimal solution:

```
-- Best Ever Individual = Individual('i', [0, 23, 12, 15, 26, 7, 22,
6, 24, 18, 10, 21, 16, 13, 17, 14, 3, 9, 19, 1, 20, 4, 28, 2, 25, 8,
11, 5, 27])

-- Best Ever Fitness = 9074.146484375
```

The solution plot is identical to the optimal one we saw previously:

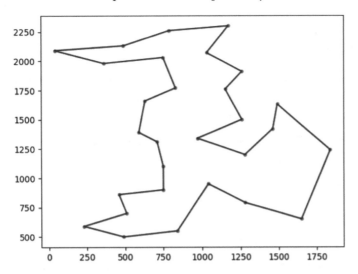

Figure 4.9: A plot of the best solution found by the program using a tournament size of 2 and elitism

The following statistics plot indicates that we were able to eliminate the "noise" we observed before. We were also able to keep some distance between the average value and the best values for a lot longer compared to the original attempt:

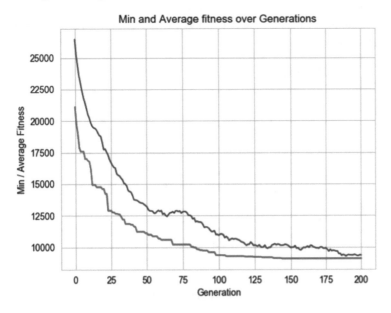

Figure 4.10: Stats of the program using a tournament size of 2 and elitism

In the next section, we will look into the **VRP**, which adds an interesting twist to the problem we just solved.

Solving the VRP

Imagine that you now manage a larger fulfillment center. You still need to deliver packages to a list of customers, but now, you have a fleet of several vehicles at your disposal. What's the best way to deliver the packages to the customers using these vehicles?

This is an example of the VRP, a generalization of the TSP described in the previous section. The basic VRP consists of the following three components:

- The list of locations that need to be visited

- The number of vehicles

- The location of the depot, which is used as the starting and ending point for each of the vehicles

The problem has numerous variations, such as several depot locations, time-critical deliveries, different types of vehicles (varying capacity, varying fuel consumption), and many more.

The goal of the problem is to minimize the cost, which can also be defined in many different ways. Examples include minimizing the time it takes to deliver all the packages, minimizing the cost of the fuel, and minimizing the variation in travel time among the vehicles used.

An illustration of a VRP with three vehicles is shown here. The cities are marked with dark circles and the depot location with an empty square, while the routes of the three vehicles are marked with three different colors:

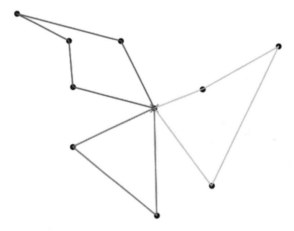

Figure 4.11: Example VRP with three vehicles

In our example, we will aim to optimize the time it takes to deliver all the packages. Since all the vehicles operate simultaneously, this measure is determined by the vehicle making the longest route. Therefore, we can make it our objective to minimize the length of the longest route among the participating vehicles' routes. For example, if we have three vehicles, each solution consists of three routes. We will evaluate all three, and then only consider the longest one of them for scoring – the longer the route, the worse the score. This will inherently encourage all three routes to be shorter, as well as closer in size to each other.

Thanks to the similarity between the two problems, we can utilize the code we wrote previously to solve the TSP for solving the VRP.

To build on the solution we created for the TSP, we can represent vehicle routing as follows:

- A TSP instance, namely a list of cities and their coordinates (or their mutual distances)
- The depot location, which is selected out of the existing cities, and represented by the index of that city
- The number of vehicles used

In the next two subsections, we will show you how to implement this solution.

Solution representation

As usual, the first question we need to address is how to represent a solution to this problem.

To illustrate our suggested representation, we will look at the 10-city example problem shown in the following figure, where the locations of the cities are marked with numbers from 0 to 9:

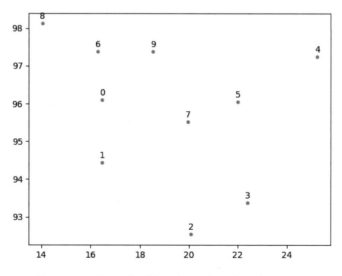

Figure 4.12: Example VRP with numbered city locations

A creative way to represent a candidate VRP solution while maintaining similarity to the previously solved TSP is by using a list that contains the numbers from 0 to $(n-1) + (m-1)$, where n is the number of cities and m is the number of vehicles. For example, if the number of cities is 10 and the number of vehicles is 3 ($n = 10$, $m = 3$), we would have a list containing all the integers from 0 to 11, as shown here:

[0, 6, 8, 9, 11, 3, 4, 5, 7, 10, 1, 2]

The first n integer values, which is 0...9 in our case, still represent the cities, just like before. However, the last $(m - 1)$ integer values, 10 and 11 in our case, are used as delimiters (or "separators") that break the list into routes. As an example, [0, 6, 8, 9 **11**, 3, 4, 5, 7, **10**, 1, 2] will be broken into the following three routes:

[0, 6, 8, 9], [3, 4, 5, 7], [1, 2]

Next, the index of the depot location needs to be removed since it is not part of a particular route. If, for example, the depot location is index **7**, the resulting routes will be as follows:

[0, 6, 8, 9], [3, 4, 5], [1, 2]

When calculating the distance that each route covers, we need to recall that each route starts and ends at the depot location (7). So, to calculate the distances, as well as plot the routes, we will be using the following data:

[7, 0, 6, 8, 9, 7], [7, 3, 4, 5, 7], [7, 1, 2, 7]

This candidate solution is illustrated in the following figure:

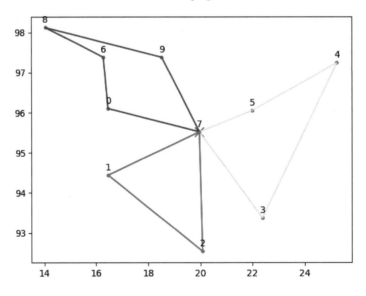

Figure 4.13: Visualization of the candidate solution [0, 6, 8, 9, 11, 3, 4, 5, 7, 10, 1, 2]

In the next subsection, we will look into a Python implementation of this idea.

Python problem representation

To encapsulate the VRP problem, we created a Python class called `VehicleRoutingProblem`. This class is contained in the `vrp.py` file and can be found at `https://github.com/PacktPublishing/Hands-On-Genetic-Algorithms-with-Python-Second-Edition/blob/main/chapter_04/vrp.py`.

The `VehicleRoutingProblem` class contains an instance of the `TravelingSalesmanProblem` class, which is used as the container for the city indices and their corresponding locations and distances. When creating an instance of the `VehicleRoutingProblem` class, the instance of the underlying `TravelingSalesmanProblem` is created internally and initialized.

The `VehicleRoutingProblem` class is initialized using the name of the underlying `TravelingSalesmanProblem`, as well as the depot location index and the number of vehicles.

In addition, The `VehicleRoutingProblem` class provides the following public methods:

- `getRoutes(indices)`: This breaks the list of given indices into separate routes by detecting the "separator" indices
- `getRouteDistance(indices)`: This calculates total the distance of the path that starts at the depot location and goes through the cities described by the given indices
- `getMaxDistance(indices)`: This calculates the max distance among the distances of the various paths described by the given indices, after breaking the indices to separate routes
- `getTotalDistance(indices)`: This calculates the combined distance of the various paths described by the given indices
- `plotData(indices)`: This breaks the list of indices into separate routes and plots each route in a different color

When executed as a standalone program, the `main` method of the class exercises these methods by creating an instance of `VehicleRoutingProblem` with the underlying TSP set to "bayg29" – the same problem we used in the previous section. The number of vehicles is set to 3, and the depot location index is set to 12 (which maps to a city with a central location). The following figure shows the locations of the cities (red dots) and the depot (green "x"):

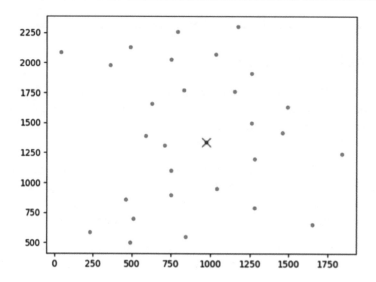

Figure 4.14: A plot of the VRP based on the "bayg29" TSP.

Red dots mark the cities while the green "X" marks the depot

The main method then generates a random solution, breaks it down into routes, and calculates the distances, as shown here:

```
random solution = [27, 23, 7, 18, 30, 14, 19, 3, 16, 2, 26, 9, 24, 22,
15, 17, 28, 11, 21, 12, 8, 4, 5, 13, 25, 6, 0, 29, 10, 1, 20]

route breakdown = [[27, 23, 7, 18], [14, 19, 3, 16, 2, 26, 9, 24, 22,
15, 17, 28, 11, 21, 8, 4, 5, 13, 25, 6, 0], [10, 1, 20]]

total distance = 26653.845703125

max distance = 21517.686
```

Note how the original list of indices of the random solution is broken down into separate routes using the separator indices (29 and 30). The plot for this random solution is shown here:

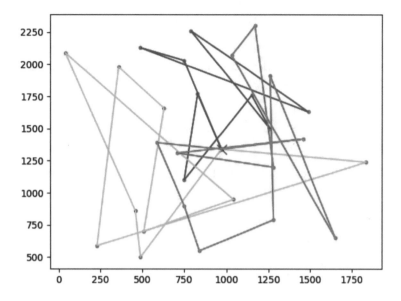

Figure 4.15: A plot of a random solution for the VRP with three vehicles

As we would expect from a random solution, it is far from optimal. This is evident from the inefficient order of cities along the long (green) route, as well as one route (green) being much longer than the other two (red and purple).

In the next subsection, we will attempt to produce good solutions using the genetic algorithms method.

Genetic algorithm solution

The genetic algorithm solution we created for the VRP resides in the 04-solve-vrp.py Python file located at https://github.com/PacktPublishing/Hands-On-Genetic-Algorithms-with-Python-Second-Edition/blob/main/chapter_04/04_solve_vrp.py.

Since we were able to build on top of the TSP and used a similar representation for the solution – an array of indices – we could use the same genetic approach we used in the previous section. We could also take advantage of elitism by reusing the elitist version that we created for the genetic flow. This makes our genetic algorithm solution very similar to the one we used for the TSP.

The following steps detail the main parts of our solution:

1. The program starts by creating an instance of the `VehicleRoutingProblem` class, using the "bayg29" TSP for its underlying data, and setting the depot location to 12 and the number of vehicles to 3:

```
TSP_NAME = "bayg29"
NUM_OF_VEHICLES = 3
DEPOT_LOCATION = 12

vrp = vrp.VehicleRoutingProblem(TSP_NAME, NUM_OF_VEHICLES,
    DEPOT_LOCATION)
```

2. The fitness function is set to minimize the distance of the longest route among the three routes produced by each solution:

```
def vrpDistance(individual):
    return vrp.getMaxDistance(individual),

toolbox.register("evaluate", vrpDistance)
```

3. For the genetic operators, we once again use tournament selection with a tournament size of 2, which is assisted by the elitist approach, and crossover and mutation operators that are specialized for ordered lists:

```
# Genetic operators:
toolbox.register("select", tools.selTournament, tournsize=2)
toolbox.register("mate", tools.cxUniformPartialyMatched, \
    indpb=2.0/len(vrp))
toolbox.register("mutate", tools.mutShuffleIndexes, \
    indpb=1.0/len(vrp))
```

4. As the VRP is inherently more difficult than TSP, we chose a larger population size and number of generations than before:

```
# Genetic Algorithm constants:
POPULATION_SIZE = 500
P_CROSSOVER = 0.9
P_MUTATION = 0.2
MAX_GENERATIONS = 1000
HALL_OF_FAME_SIZE = 30
```

And that's it! We're ready to run the program. The results that we obtain with these settings are shown here – three routes, with a maximum length of 3857:

```
-- Best Ever Individual =   Individual('i', [0, 20, 17, 16, 13, 21, 10,
14, 3, 29, 15, 23, 7, 26, 12, 22, 6, 24, 18, 9, 19, 30, 27, 11, 5, 4,
8, 25, 2, 28, 1])

-- Best Ever Fitness =   3857.36376953125

-- Route Breakdown =   [[0, 20, 17, 16, 13, 21, 10, 14, 3], [15, 23, 7,
26, 22, 6, 24, 18, 9, 19], [27, 11, 5, 4, 8, 25, 2, 28, 1]]

-- total distance =   11541.875

-- max distance =   3857.3638
```

Note, again, how the solution is broken down into three separate routes using the highest two indices (29, 30) as separators, and ignoring the depot location (12). We ended up with three routes, two of them covering nine cities each, and the third covering 10 cities.

Plotting the solution produces the following figure depicting the three resulting routes:

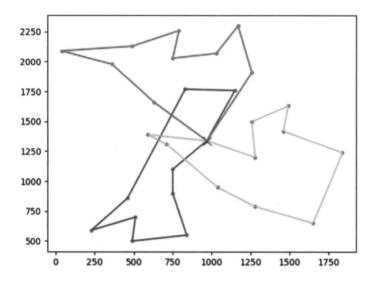

Figure 4.16: A plot of the best solution found by the program for the VRP with three vehicles

The following statistics plot shows that the algorithm did most of the optimization before reaching 300 generations. After, there are several small improvements:

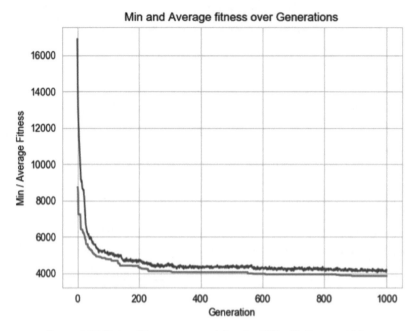

Figure 4.17: Stats of the program solving the VRP with three vehicles

How about changing the number of vehicles? Let's run the algorithm again after increasing the number of vehicles to six, and making no other changes:

```
NUM_OF_VEHICLES = 6
```

The results for this run are shown here – six routes, with a maximum length of 2803:

```
-- Best Ever Individual =  Individual('i', [27, 11, 5, 8, 4, 33, 12,
24, 6, 22, 7, 23, 29, 28, 20, 0, 26, 15, 32, 3, 18, 13, 17, 1, 31, 19,
25, 2, 30, 9, 14, 16, 21, 10])


-- Best Ever Fitness =  2803.584716796875


-- Route Breakdown =  [[27, 11, 5, 8, 4], [24, 6, 22, 7, 23], [28, 20,
0, 26, 15], [3, 18, 13, 17, 1], [19, 25, 2], [9, 14, 16, 21, 10]]


-- total distance =  16317.9892578125


-- max distance =  2803.5847
```

Note that increasing the number of vehicles two-fold didn't decrease the maximum distance in a similar manner (2803 with six compared to 3857 with three). This is likely because each separate route still needs to start and end at the depot location, which is added to the cities in the route.

Plotting the solution produces the following figure, depicting the six resulting routes:

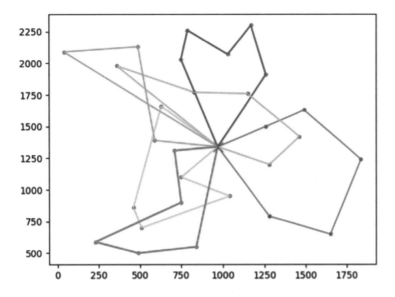

Figure 4.18: A plot of the best solution found by the program for the VRP with six vehicles

One interesting point demonstrated by this plot is that the orange route doesn't seem to be optimized. Since we told the genetic algorithm to minimize the longest route, any route that is shorter than the longest route may not be optimized further. You are encouraged to modify our solution to further optimize the routes.

As with the three-vehicle case, the following statistics plot shows that the algorithm did most of the optimization before reaching 200 generations, after which there are several small improvements:

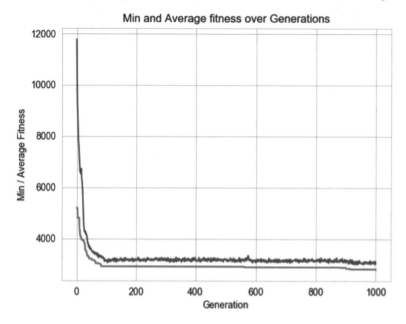

Figure 4.19: Stats of the program solving the VRP with six vehicles

The solution we've found seems reasonable. Can we do better than that? What about other numbers of vehicles? Or other depot locations? Different genetic operators or different parameter settings? Perhaps even a different fitness criteria? We encourage you to experiment with all these and learn from these experiments.

Summary

In this chapter, you were introduced to search problems and combinatorial optimization. We then closely examined three classic combinatorial problems – each with numerous real-life applications – the knapsack problem, the TSP, and the VRP. For Each of these problems, we followed a similar process of finding an appropriate representation for a solution, creating a class that encapsulates the problem and evaluates a given solution, and then creating a genetic algorithm solution that utilizes that class. We ended up with valid solutions for all three problems while experimenting with genotype-to-phenotype mapping and elitism-backed exploration.

In the next chapter, we will look into a family of closely related tasks, namely *constraint satisfaction problems*, starting with the classic *n-queens* problem.

Further reading

For more information, please refer to the following resources:

- Solving the knapsack problem using dynamic programming, from the book *Keras Reinforcement Learning Projects*, by Giuseppe Ciaburro, September 2018

- The VRP, from the book *Keras Reinforcement Learning Projects*, by Giuseppe Ciaburro, September 2018

5

Constraint Satisfaction

In this chapter, you will learn how genetic algorithms can be utilized to solve constraint satisfaction problems. We will start by describing the concept of constraint satisfaction and how it applies to search problems and combinatorial optimization. Then, we will look at several hands-on examples of constraint satisfaction problems and their Python-based solutions using the DEAP framework. The problems we will cover include the well-known **N-Queen** problem, followed by the **nurse scheduling** problem, and finally the **graph coloring** problem. Along the way, we will learn the difference between **hard and soft constraints**, as well as how they can be incorporated into the solution process.

In this chapter, we will cover the following topics:

- Understanding the nature of constraint satisfaction problems
- Solving the N-Queens problem using a genetic algorithm coded with the DEAP framework
- Solving an example of the nurse scheduling problem using a genetic algorithm coded with the DEAP framework
- Solving the graph coloring problem using a genetic algorithm coded with the DEAP framework
- Understanding the concepts of hard and soft constraints, as well as how to apply them when solving a problem

Technical requirements

In this chapter, we will be using Python 3 with the following supporting libraries:

- `deap`
- `numpy`
- `matplotlib`
- `seaborn`
- `networkx` – introduced in this chapter

> **Important note**
>
> If you're using the `requirements.txt` file we've provided (see *Chapter 3*), these libraries will already be in your environment.

The programs that will be used in this chapter can be found in this book's GitHub repository at `https://github.com/PacktPublishing/Hands-On-Genetic-Algorithms-with-Python-Second-Edition/tree/main/chapter_05`.

Check out the following video to see the Code in Action: `https://packt.link/OEBOd`.

Constraint satisfaction in search problems

In the previous chapter, we looked at solving search problems, which focused on methodically evaluating states and transitions between states. Every state transition typically involves a cost or gain, and the objective of the search was to minimize the cost or maximize the gain. Constraint satisfaction problems are a *variant* of search problems, where the states must satisfy several constraints or limitations. If we can translate the various violations of constraints into cost and then strive to minimize the cost, solving a constraint satisfaction problem can resemble solving a general search problem.

Like combinatorial optimization problems, constraint satisfaction problems have important applications in fields such as artificial intelligence, operations research, and pattern matching. A better understanding of these problems may help in solving numerous types of problems that may seem unrelated at first glance. Constraint satisfaction problems often exhibit high complexity, which makes genetic algorithms a suitable candidate for solving them.

The **N-Queens** problem, which will be presented in the next section, illustrates the concept of constraint satisfaction problems and demonstrates how they can be solved in a very similar manner to the problems we looked at in the previous chapter.

Solving the N-Queens problem

Originally known as the *eight-queen puzzle*, the classic N-Queens problem originated from the game of chess, and the *8x8* chessboard was its early playground. The task was to place eight chess queens on the board without any two of them threatening each other. In other words, no two queens can share the same row, same column, or same diagonal. The N-Queens problem is similar, using an *N×N* chessboard and *N* chess queens.

The problem is known to have a solution for any natural number, *n*, except for the cases of *n=2* and *n=3*. For the original eight-queen case, there are 92 solutions, or 12 unique solutions if we consider symmetrical solutions to be identical. One of the solutions is as follows:

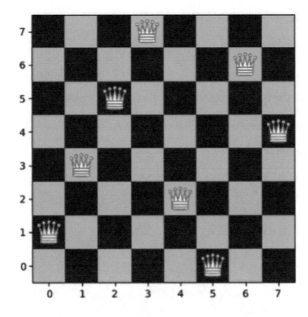

Figure 5.1: One of the 92 possible solutions for the eight-queen puzzle

By applying combinatorics, the count of all possible ways to place eight pieces on the *8×8* board yields 4,426,165,368 combinations. However, if we can create our candidate solutions in a way that ensures that no two queens will be placed on the same row or the same column, the number of possible combinations is dramatically reduced to *8!* (factorial of 8), which amounts to 40,320. We are going to take advantage of this idea in the next subsection when we choose the way our solution to this problem will be represented.

Solution representation

When solving the N-Queens problem, we can take advantage of the knowledge that each row will host exactly one queen, and no two queens will share the same column. This means we can represent any candidate solution as an ordered list of integers – or a list of indices, with each index representing the column that one of the queens occupies for the current row.

For example, in a four-queen problem over a 4×4 chessboard, we have the following list of indices:

```
[3, 2, 0, 1]
```

This translates to the following positions:

- In the first row, the queen is placed in position 3 (fourth column)
- In the second row, the queen is placed in position 2 (third column).
- In the third row, the queen is placed in position 0 (first column)
- In the fourth row, the queen is placed in position 1 (second column)

This is depicted in the following figure:

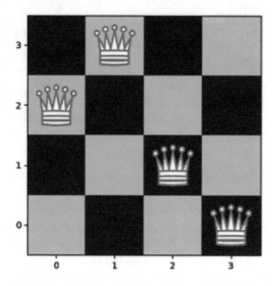

Figure 5.2: Illustration of the queen arrangement represented by the list [3, 2, 0, 1]

Similarly, another arrangement of the indices may look as follows:

```
[1, 3, 0, 2]
```

This arrangement represents the candidate solution shown in the following figure:

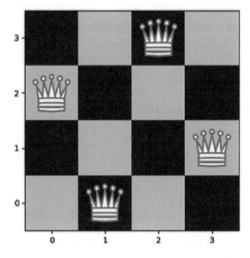

Figure 5.3: Illustration of the queen arrangement represented by the list [1, 3, 0, 2]

The only constraint violations that are possible in candidate solutions represented this way are shared diagonals between pairs of queens.

For example, the first candidate solution we discussed contains two violations, as shown here:

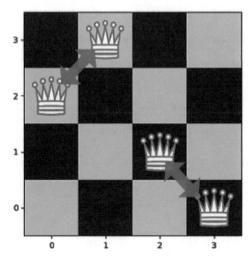

Figure 5.4: Illustration of the queen arrangement represented by the
list [3, 2, 0, 1], with constraint violations indicated

However, the preceding one exhibited no violations.

This means that, when evaluating the solutions that are represented in this way, we only need to find and count the shared diagonals between the positions they stand for.

The solution representation we just discussed is a central part of the Python class that we will describe in the next subsection.

Python problem representation

To encapsulate the N-Queens problem, we've created a Python class called NQueensProblem. This class can be found in the queens.py file of this book's GitHub repository: https://github. com/PacktPublishing/Hands-On-Genetic-Algorithms-with-Python-Second-Edition/blob/main/chapter_05/queens.py.

The class is initialized with the desired size of the problem and provides the following public methods:

- getViolationsCount(positions): This calculates the number of violations in the given solution, which is represented by a list of indices, as discussed in the previous subsection
- plotBoard(positions): This plots the positions of the queens on the board according to the given solution

The main method of the class exercises the class methods by creating an eight-queen problem and testing the following candidate solution for it:

```
[1, 2, 7, 5, 0, 3, 4, 6]
```

This is followed by plotting the candidate solution and calculating the number of constraint violations.

The resulting output is shown here:

```
Number of violations =  3
```

The plot for this is as follows – can you spot all three violations?

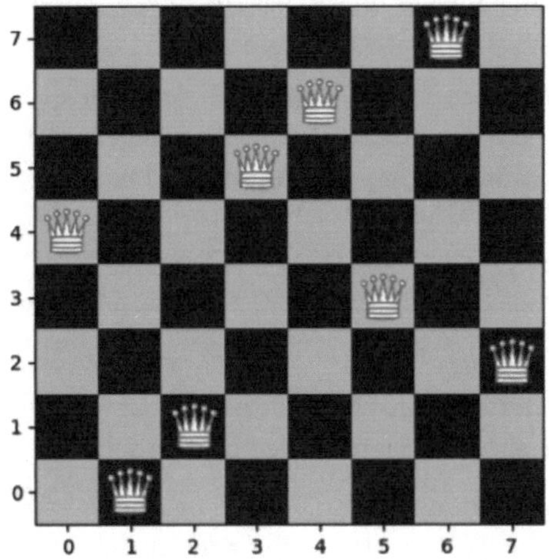

Figure 5.5: Illustration of the eight-queen arrangement represented by the list [1, 2, 7, 5, 0, 3, 4, 6]

In the next subsection, we will apply the genetic algorithm approach to solving the N- Queens problem.

Genetic algorithms solution

To solve the N-Queens problem using a genetic algorithm, we've created a Python program called `01-solve-n-queens.py`, which is located at `https://github.com/PacktPublishing/Hands-On-Genetic-Algorithms-with-Python-Second-Edition/blob/main/chapter_05/01_solve_n_queens.py`.

Since the solution representation we chose for this problem is a list (or an array) of indices, which is similar to the representation we used for the **traveling salesman problem** (**TSP**) and the **vehicle routing problem** (**VRP**) in *Chapter 4, Combinatorial Optimization*, we can utilize a similar genetic approach to the one we used there. In addition, we'll take advantage of **elitism** once more by reusing the elitist version that we created for DEAP's simple genetic flow.

The following steps describe the main parts of our solution:

1. Our program starts by creating an instance of the `NQueensProblem` class using the size of the problem we would like to solve:

    ```
    nQueens = queens.NQueensProblem(NUM_OF_QUEENS)
    ```

2. Since our goal is to minimize the count of violations (hopefully to a value of 0), we define a single objective, minimizing the fitness strategy:

    ```
    creator.create("FitnessMin", base.Fitness, weights=(-1.0,))
    ```

3. Since the solution is represented by an ordered list of integers, where each integer denotes the column location of a queen, we can use the following toolbox definitions to create the initial population:

```
# create an operator that generates randomly shuffled indices:
toolbox.register("randomOrder", random.sample, \
    range(len(nQueens)), len(nQueens))
toolbox.register("individualCreator", tools.initIterate, \
    creator.Individual, toolbox.randomOrder)
toolbox.register("populationCreator", tools.initRepeat, \
    list, toolbox.individualCreator)
```

4. The actual fitness function is set to count the number of violations caused by the placement of the queens on the chessboard, as represented by each solution:

```
def getViolationsCount(individual):
    return nQueens.getViolationsCount(individual),
toolbox.register("evaluate", getViolationsCount)
```

5. As for the genetic operators, we use *tournament selection* with a tournament size of 2, as well as the *crossover* and *mutation* operators, which are specialized for ordered lists:

```
# Genetic operators:
toolbox.register("select", tools.selTournament, \
    tournsize=2)
toolbox.register("mate", tools.cxUniformPartialyMatched, \
    indpb=2.0/len(nQueens))
toolbox.register("mutate", tools.mutShuffleIndexes, \
    indpb=1.0/len(nQueens))
```

6. In addition, we continue to use the **elitist** approach, where the **hall-of-fame (HOF)** members – the current best individuals – are always passed untouched to the next generation. As we found out in the previous chapter, this approach works well with a tournament selection of size 2:

```
population, logbook = elitism.eaSimpleWithElitism(population,
    toolbox,
    cxpb=P_CROSSOVER,
    mutpb=P_MUTATION,
    ngen=MAX_GENERATIONS,
    stats=stats,
    halloffame=hof,
    verbose=True)
```

7. Since each N-Queens problem can have multiple possible solutions, we print out all HOF members, instead of just the top one, so that we can see how many valid solutions we found:

```
print("- Best solutions are:")
for i in range(HALL_OF_FAME_SIZE):
    print(i, ": ", hof.items[i].fitness.values[0], " -> ",
        hof.items[i])
```

As we saw earlier, our solution representation reduces the eight-queen case to only about 40,000 possible combinations, which makes it a rather small problem. To make things more interesting, let's increase the size to 16 queens, where the number of possible candidate solutions will be *16!*. This calculates to the colossal value of 20,922,789,888,000. The number of valid solutions to this problem is quite large too, at just under 15 million. However, compared to the number of possible combinations, searching for a valid solution is still like trying to find a needle in a haystack.

Before we run the program, let's set the algorithm constants, as follows:

```
NUM_OF_QUEENS = 16
POPULATION_SIZE = 300
MAX_GENERATIONS = 100
HALL_OF_FAME_SIZE = 30
P_CROSSOVER = 0.9
P_MUTATION = 0.1
```

Running the program with these settings yields the following output:

```
gen nevals min avg
0 300 3 10.4533
1 246 3 8.85333
..
23 250 1 4.38
24 227 0 4.32
..
- Best solutions are:
0 : 0.0 -> Individual('i', [7, 2, 8, 14, 9, 4, 0, 15, 6, 11, 13, 1, 3,
5, 10, 12])
1 : 0.0 -> Individual('i', [7, 2, 6, 14, 9, 4, 0, 15, 8, 11, 13, 1, 3,
5, 12, 10])
..
7 : 0.0 -> Individual('i', [14, 2, 6, 12, 7, 4, 0, 15, 8, 11, 3, 1, 9,
5, 10, 13])
8 : 1.0 -> Individual('i', [2, 13, 6, 12, 7, 4, 0, 15, 8, 14, 3, 1, 9,
5, 10, 11])
..
```

From the printouts, we can see that a solution was first found in generation 24, where the fitness value shows as 0, which means no violations. In addition, the printout of the best solutions indicates that eight different solutions were found during the run. These solutions are the entries of 0 to 7 in the HOF, all of which have a fitness value of 0. The next entry already has a fitness value of 1, denoting a violation.

The first plot that's produced by the program depicts the placement of the 16 queens on the *16x16* chessboard, as defined by the first valid solution that was found – [7, 2, 8, 14, 9, 4, 0, 15, 6, 11, 13, 1, 3, 5, 10, 12]:

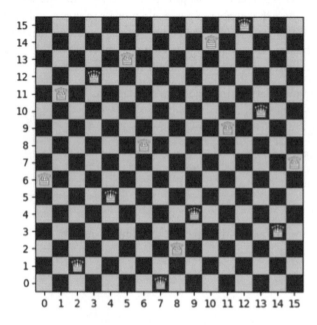

Figure 5.6: A plot of a valid 16-queen arrangement found by the program

The second plot contains a graph of the max and average fitness values over the generations. From this graph, we can see that even though the best fitness value of zero was found early on – around generation 24 – the average fitness value kept decreasing as more solutions were found:

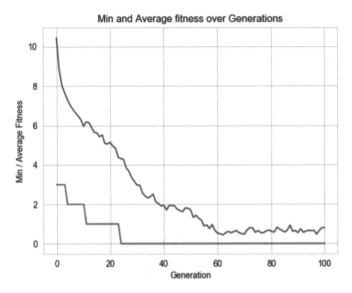

Figure 5.7: Stats of the program solving the 16-queen problem

Increasing the value of MAX_GENERATIONS to 400 without making any other changes will result in us finding 38 valid solutions. If we increase MAX_GENERATIONS to 500, all 50 members of the HOF will contain valid solutions. You are encouraged to try out various combinations of the genetic algorithm's settings, as well as solve other sizes of the N-Queen problem.

In the next section, we will be transitioning from arranging game pieces on a board to placing workers on a work schedule.

Solving the nurse scheduling problem

Imagine that you are responsible for scheduling the shifts for the nurses in your hospital department for this week. There are three shifts in a day – morning, afternoon, and night – and for each shift, you need to assign one or more of the eight nurses who work in your department. If this sounds like a simple task, take a look at the list of relevant hospital rules:

- A nurse is not allowed to work two consecutive shifts

- A nurse is not allowed to work more than five shifts per week

- The number of nurses per shift in your department should fall within the following limits:

  - Morning shift: 2-3 nurses

  - Afternoon shift: 2-4 nurses

  - Night shift: 1-2 nurses

In addition, each nurse can have shift preferences. For example, one nurse prefers to only work morning shifts, another nurse prefers to not work afternoon shifts, and so on.

This task is an example of the **nurse scheduling problem** (**NSP**), which can have many variants. Possible variations may include different specialties for different nurses, the ability to work on cover shifts (overtime), or even different types of shifts – such as 8-hour shifts and 12-hour shifts.

By now, it probably seems like a good idea to write a program that will do the scheduling for you. Why not apply our knowledge of genetic algorithms to implement such a program? As usual, we will start by representing the solution to the problem.

Solution representation

To solve the nurse scheduling problem, we decided to use a binary list (or array) to represent the schedule as it will be intuitive for us to interpret, and we've seen that genetic algorithms can handle this representation naturally.

For each nurse, we can have a **binary string** representing the 21 shifts of the week. A value of 1 represents a shift that the nurse is scheduled to work on. For example, take a look at the following binary list:

```
[0, 1, 0, 1, 0, 1, 0, 1, 1, 0, 0, 0, 0, 0, 1, 1, 0, 0, 0, 1, 0]
```

This list can be broken into the following groups of three values, representing the shifts this nurse will be working each day of the week:

Sunday	Monday	Tuesday	Wednesday	Thursday	Friday	Saturday
[0, 1, 0]	[1, 0, 1]	[0, 1, 1]	[0, 0, 0]	[0, 0, 1]	[1, 0, 0]	[0, 1, 0]
Afternoon	Morning and night	Afternoon and night	None	Night	Morning	Afternoon

Table 5.1: Converting the binary sequence into daily shifts

The schedules of all nurses can be then concatenated together to create one long binary list representing the entire solution.

When evaluating a solution, this long list can be broken down into the schedules of the individual nurses, and violations of the constraints can be checked for. The preceding sample nurse schedule, for instance, contains two occurrences of consecutive 1 values that represent consecutive shifts being worked (afternoon followed by night, and night followed by morning). The number of weekly shifts for that same nurse can be calculated by totaling the binary values of the list, which results in 8 shifts. We can also easily check for adherence to the shift preferences by checking each day's shifts against the given preferred shifts of that nurse.

Finally, to check for the constraints of the number of nurses per shift, we can sum the weekly schedules of all nurses and look for entries that are larger than the maximum allowed or smaller than the minimum allowed.

But before we continue with our implementation, we need to discuss the difference between **hard constraints** and **soft constraints**.

Hard constraints versus soft constraints

When solving the nurse scheduling problem, we should bear in mind that some of the constraints represent hospital rules that cannot be broken. A schedule that contains one or more violations of these rules will be considered invalid. More generally, these are known as **hard constraints**.

The nurses' preferences, on the other hand, can be considered **soft constraints**. We would like to adhere to them as much as possible, and a solution that contains no violations or fewer violations of these constraints is considered better than one that contains more violations. However, a violation of these constraints does not invalidate the solution.

In the case of the N-Queens problem, all the constraints – row, column, and diagonal – were hard constraints. Had we not found a solution where the number of violations was zero, we would not have a valid solution for the problem. Here, on the other hand, we are looking for a solution that will not violate any of the hospital rules while minimizing the number of breaches to the nurses' preferences.

While dealing with soft constraints is similar to what we do in any optimization problem – that is, we strive to minimize them – how do we deal with the hard constraints that accompany them? There are several possible strategies:

- Find a particular representation (coding) of the solution that **eliminates the possibility** of a hard constraint violation. When solving the N-Queens problem, we were able to represent a solution in a way that eliminated the possibility for two of the three constraints – row and column – which considerably simplified our solution. But generally, such coding may be difficult to find.

- When evaluating the solutions, **discard** candidate solutions that violate any hard constraint. The disadvantage of this approach is the loss of information contained in these solutions, which may be valuable for the problem. This could considerably slow down the optimization process.

- When evaluating the solutions, **repair** candidate solutions that violate any hard constraint. In other words, find a way to manipulate the solution and modify it so that it will no longer violate the constraint(s). Creating such a repair procedure can prove difficult or impossible for most problems, and at the same time, the repair process may result in a significant loss of information.

- When evaluating the solutions, **penalize** candidate solutions that violate any hard constraint. This will degrade the solution's score and make it less desirable, but will not eliminate it completely, so the information contained in it is not lost. Effectively, this leads to a hard constraint to be considered similar to a soft constraint, but with a heavier penalty. When using this method, the challenge may be to find the appropriate extent of the penalty. Too harsh a penalty may lead to a de facto elimination of such solutions, while a penalty that's too small may lead to these solutions appearing as optimal.

In our case, we chose to apply the fourth approach and penalize the violations of the hard constraints to a larger degree than those of the soft constraints. We did this by creating a cost function, where the cost of a hard constraint violation is greater than that of a soft constraint violation. The total cost is then used as the fitness function to be minimized. This is implemented within the problem representation that will be discussed in the next subsection.

Python problem representation

To encapsulate the nurse scheduling problem we described at the beginning of this section, we've created a Python class called `NurseSchedulingProblem`. This class is contained in the `nurses.py` file, which can be found at `https://github.com/PacktPublishing/Hands-On-Genetic-Algorithms-with-Python-Second-Edition/blob/main/chapter_05/nurses.py`.

The class constructor accepts the `hardConstraintPenalty` parameter, which represents the penalty factor for a hard constraint violation (while the penalty of a soft constraint violation is fixed to 1). Then, it continues to initialize the various parameters, describing the scheduling problem:

```
# list of nurses:
self.nurses = ['A', 'B', 'C', 'D', 'E', 'F', 'G', 'H']
# nurses' respective shift preferences - morning, evening, night:
self.shiftPreference = [[1, 0, 0], [1, 1, 0], [0, 1, 1], [0, 1, 0],
    [0, 0, 1], [1, 1, 1], [0, 1, 1], [1, 1, 1]]
# min and max number of nurses allowed for each shift - morning,
evening, night:
self.shiftMin = [2, 2, 1]
self.shiftMax = [3, 4, 2]
# max shifts per week allowed for each nurse:
self.maxShiftsPerWeek = 5
```

The class uses the following method to convert the given schedule into a dictionary with a separate schedule for each nurse:

- `getNurseShifts(schedule)`

The following methods are used to count the various types of violations:

- `countConsecutiveShiftViolations(nurseShiftsDict)`
- `countShiftsPerWeekViolations(nurseShiftsDict)`
- `countNursesPerShiftViolations(nurseShiftsDict)`
- `countShiftPreferenceViolations(nurseShiftsDict)`

In addition, the class provides the following public methods:

- `getCost(schedule)`: Calculates the total cost of the various violations in the given schedule. This method uses the value of the `hardConstraintPenalty` variable.
- `printScheduleInfo(schedule)`: Prints the schedule and violation details.

The main method of the class exercises the class' methods by creating an instance of the nurse scheduling problem and testing a randomly generated solution for it. The resulting output may look as follows, with the value of `hardConstraintPenalty` set to `10`:

```
Random Solution =
[0 1 0 0 0 1 0 0 0 1 0 0 0 0 1 0 1 1 1 0 1 0 1 1 1 1 1 1 1 0 0 1 1 1
 0 1 0 0 0 0 0 1 1 1 1 0 1 1 0 1 0 1 0 1 1 0 0 0 0 0 0 0 1 1 0 1 1
 1 1 0 1 0 1 1 1 0 1 0 1 0 1 0 0 1 0 1 1 1 1 1 1 1 1 1 0 0 1 1 1 1
 1 1 1 1 0 1 0 1 1 0 1 0 1 1 0 1 0 1 0 0 1 1 0 1 1 1 0 0 0 0 0 0 0 0
 1 0 1 1 1 0 0 0 0 1 0 0 0 0 0 1 0 1 0 1 0 0 1 1 1 0 1 0]
Schedule for each nurse:
A : [0 1 0 0 0 1 0 0 0 1 0 0 0 0 1 0 1 1 1 0 1]
B : [0 1 1 1 1 1 1 1 1 0 0 1 1 1 0 1 0 0 0 0 0]
C : [1 1 1 1 1 0 1 1 0 1 0 1 0 1 0 1 1 0 0 0 0 0]
D : [0 0 1 1 0 1 1 1 1 0 1 0 1 1 1 0 1 0 1 0 1]
E : [0 0 1 0 1 1 1 1 1 1 1 1 1 1 0 0 1 1 1 1]
F : [1 1 1 1 0 1 0 1 1 0 1 0 1 1 0 1 0 1 0 0 1]
G : [1 0 1 1 1 0 0 0 0 0 0 0 0 1 0 1 1 1 0 0]
H : [0 0 1 0 0 0 0 0 1 0 1 0 1 0 0 1 1 1 0 1 0]
consecutive shift violations =  47
weekly Shifts =  [8, 12, 11, 13, 16, 13, 8, 8]
Shifts Per Week Violations =  49
Nurses Per Shift =  [3, 4, 7, 5, 4, 5, 4, 5, 5, 3, 4, 3, 5, 5, 5, 3,
4, 5, 4, 2, 4]
Nurses Per Shift Violations =  28
Shift Preference Violations =  39
Total Cost =  1279
```

As is evident from these results, a randomly generated solution is likely to yield a large number of violations, and consequently a large cost value. In the next subsection, we'll attempt to minimize the cost and eliminate all hard constraint violations using a genetic algorithm-based solution.

Genetic algorithms solution

To solve the nurse scheduling problem using a genetic algorithm, we've created a Python program called `02-solve-nurses.py`, which is located at `https://github.com/PacktPublishing/Hands-On-Genetic-Algorithms-with-Python-Second-Edition/blob/main/chapter_05/02_solve_nurses.py`.

Since the solution representation we chose for this problem is a list (or an array) of binary values, we were able to use the same genetic approach we used for several problems we have solved already, such as the 0-1 knapsack problem we described in *Chapter 4, Combinatorial Optimization*.

The main parts of our solution are described in the following steps:

1. Our program starts by creating an instance of the `NurseSchedulingProblem` class with the desired value for `hardConstraintPenalty`, which is set by the HARD_CONSTRAINT_PENALTY constant:

    ```
    nsp = nurses.NurseSchedulingProblem(HARD_CONSTRAINT_PENALTY)
    ```

2. Since our goal is to minimize the cost, we must define a single objective, minimizing the fitness strategy:

    ```
    creator.create("FitnessMin", base.Fitness, weights=(-1.0,))
    ```

3. Since the solution is represented by a list of 0 or 1 values, we must use the following toolbox definitions to create the initial population:

    ```
    creator.create("Individual", list, fitness=creator.FitnessMin)
    toolbox.register("zeroOrOne", random.randint, 0, 1)
    toolbox.register("individualCreator", tools.initRepeat, \
        creator.Individual, toolbox.zeroOrOne, len(nsp))
    toolbox.register("populationCreator", tools.initRepeat, \
        list, toolbox.individualCreator)
    ```

4. The actual fitness function is set to calculate the cost of the various violations in the schedule, represented by each solution:

    ```
    def getCost(individual):
        return nsp.getCost(individual),
    toolbox.register("evaluate", getCost)
    ```

5. As for the genetic operators, we must use tournament selection with a tournament size of 2, along with two-point crossover and flip-bit mutation, since this is suitable for binary lists:

    ```
    toolbox.register("select", tools.selTournament, tournsize=2)
    toolbox.register("mate", tools.cxTwoPoint)
    toolbox.register("mutate", tools.mutFlipBit, indpb=1.0/len(nsp))
    ```

6. We keep using the elitist approach, where HOF members – the current best individuals – are always passed untouched to the next generation:

```
population, logbook = elitism.eaSimpleWithElitism(
    population, toolbox, cxpb=P_CROSSOVER, \
    mutpb=P_MUTATION, ngen=MAX_GENERATIONS, \
    stats=stats, halloffame=hof, verbose=True)
```

7. When the algorithm concludes, we print the details of the best solution that was found:

```
nsp.printScheduleInfo(best)
```

Before we run the program, let's set the algorithm constants, as follows:

```
POPULATION_SIZE = 300
P_CROSSOVER = 0.9
P_MUTATION = 0.1
MAX_GENERATIONS = 200
HALL_OF_FAME_SIZE = 30
```

In addition, let's start by setting the penalty for violating hard constraints to a value of 1, which makes the cost of violating a hard constraint similar to that of violating a soft constraint:

```
HARD_CONSTRAINT_PENALTY = 1
```

Running the program with these settings yields the following output:

```
-- Best Fitness = 3.0
-- Schedule =
Schedule for each nurse:
A : [0, 0, 0, 0, 0, 0, 1, 0, 0, 1, 0, 0, 1, 0, 0, 1, 0, 0, 1, 0, 0]
B : [1, 0, 0, 1, 0, 0, 1, 0, 0, 1, 0, 0, 0, 1, 0, 0, 0, 0, 1, 0, 0]
C : [0, 1, 0, 0, 0, 0, 0, 0, 0, 0, 0, 1, 0, 0, 0, 0, 0, 0, 0, 0, 0]
D : [0, 1, 0, 0, 0, 0, 0, 0, 0, 0, 1, 0, 0, 1, 0, 0, 1, 0, 0, 1, 0]
E : [0, 0, 1, 0, 1, 0, 0, 0, 0, 0, 0, 0, 0, 0, 1, 0, 0, 1, 0, 0, 0]
F : [0, 0, 0, 0, 0, 1, 0, 1, 0, 0, 1, 0, 0, 0, 0, 0, 0, 1, 0, 1, 0]
G : [0, 0, 0, 0, 1, 0, 0, 0, 1, 0, 0, 1, 0, 0, 0, 0, 1, 0, 0, 0, 1]
H : [1, 0, 0, 1, 0, 0, 0, 1, 0, 0, 0, 0, 1, 0, 0, 1, 0, 0, 0, 0, 0]
consecutive shift violations = 0
weekly Shifts = [5, 6, 2, 5, 4, 5, 5, 5]
Shifts Per Week Violations = 1
Nurses Per Shift = [2, 2, 1, 2, 2, 1, 2, 2, 1, 2, 2, 2, 2, 2, 1, 2, 2,
2, 2, 2, 1]
Nurses Per Shift Violations = 0
Shift Preference Violations = 2
```

This may seem like a good result since we ended up with only three constraint violations. However, one of them is a **shift-per-week violation** – nurse B was scheduled with six shifts for the week, exceeding the maximum allowed of five. This is enough to make the entire solution unacceptable.

In an attempt to eliminate this kind of violation, we'll proceed to increase the hard constraint penalty value to `10`:

```
HARD_CONSTRAINT_PENALTY = 10
```

Now, the result is as follows:

```
-- Best Fitness = 3.0
-- Schedule =
Schedule for each nurse:
A : [0, 0, 0, 1, 0, 0, 1, 0, 0, 1, 0, 0, 0, 0, 0, 0, 0, 1, 0, 0]
B : [1, 0, 0, 1, 0, 0, 0, 0, 0, 1, 0, 0, 1, 0, 0, 1, 0, 0, 0, 0]
C : [0, 0, 1, 0, 0, 0, 0, 0, 1, 0, 0, 0, 0, 0, 1, 0, 0, 1, 0, 0, 1]
D : [0, 1, 0, 0, 0, 0, 0, 0, 0, 0, 1, 0, 0, 1, 0, 0, 1, 0, 0, 1, 0]
E : [0, 0, 0, 0, 1, 0, 0, 1, 0, 0, 0, 0, 0, 1, 0, 0, 0, 0, 0, 0, 0]
F : [0, 0, 0, 0, 0, 0, 1, 0, 0, 0, 1, 0, 1, 0, 0, 0, 1, 0, 1, 0, 0]
G : [0, 1, 0, 0, 1, 0, 0, 0, 1, 0, 0, 1, 0, 0, 0, 0, 0, 1, 0, 0, 0]
H : [1, 0, 0, 0, 0, 1, 0, 1, 0, 0, 0, 0, 0, 0, 0, 1, 0, 0, 0, 1, 0]
consecutive shift violations = 0
weekly Shifts = [4, 5, 5, 5, 3, 5, 5, 5]
Shifts Per Week Violations = 0
Nurses Per Shift = [2, 2, 1, 2, 2, 1, 2, 2, 2, 2, 2, 1, 2, 2, 1, 2, 2,
2, 2, 2, 1]
Nurses Per Shift Violations = 0
Shift Preference Violations = 3
```

Again, we got three violations, but this time, they were all soft constraint violations, which makes this solution valid.

The following graph, which depicts the minimum and average fitness over the generations, indicates that over the first 40-50 generations, the algorithm was able to eliminate all hard constraint violations, and from there on there were only small incremental improvements, which occurred whenever another soft constraint was eliminated:

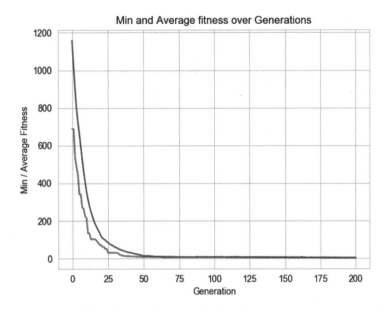

Figure 5.8: Stats of the program solving the nurse scheduling problem

It seems that, in our case, it was enough to set a 10-fold penalty on hard constraint violations. In other problems, higher values may be required. You are encouraged to experiment by altering the problem's definitions, as well as the genetic algorithm's settings.

The same trade-off we have just seen between soft and hard constraints is going to play a part in the next task we take on – the graph coloring problem.

Solving the graph coloring problem

In the mathematical branch of graph theory, a **graph** is a structured collection of objects that represents the relationships between pairs of these objects. The objects appear as **vertices** (or nodes) in the graph, while the relation between a pair of objects is represented using an **edge**. A common way of illustrating a graph is by drawing the vertices as circles and the edges as connecting lines, as depicted in the following diagram of the *Petersen graph*, named after the Danish mathematician Julius Petersen:

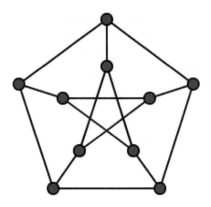

Figure 5.9: Petersen graph

Source: https://commons.wikimedia.org/wiki/File:Petersen1_tiny.svg

Image by Leshabirukov.

Graphs are remarkably useful objects as they can represent and help us research an overwhelming variety of real-life structures, patterns, and relationships, such as social networks, power grid layouts, website structures, linguistic compositions, computer networks, atomic structures, migration patterns, and more.

The **graph coloring** task is used to assign a color for every node in the graph in such a way that no pair of connected (adjacent) nodes will share the same color. This is also known as the **proper coloring** of the graph.

The following diagram shows the same Petersen graph, but this time colored properly:

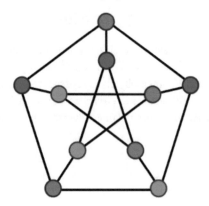

Figure 5.10: Proper coloring of the Petersen graph

Source: https://en.wikipedia.org/wiki/File:Petersen_graph_3-coloring.svg.

The color assignment is often accompanied by an optimization requirement – use the **minimum possible** number of colors. For example, the Peterson graph can be properly colored using three colors, as demonstrated in the preceding diagram. But it would be impossible to color it properly using only two colors. In graph theory terms, this means that the *chromatic number* of this graph is three.

Why would we care about coloring the nodes of the graph? Many real-life problems can be translated into a graph representation in such a way that the graph coloring will stand for a solution – for example, scheduling classes for a student, or shifts for an employee can be translated into a graph, where adjacent nodes represent classes or shifts that cause a conflict. Such a conflict can be classes that fall at the same time or shifts that are consecutive (sound familiar?). Due to this conflict, assigning the same person to both classes (or both shifts) will invalidate the schedule. If each color represents a different person, assigning different colors to adjacent nodes will solve the conflicts. The N-Queen problem we encountered at the beginning of this chapter can be represented as a graph coloring problem, where every node in the graph represents a square on the chessboard, and every pair of nodes that share a row, a column, or a diagonal is connected by an edge. Other relevant examples include frequency assignments to radio stations, power grid redundancy planning, traffic light timing, and even Sudoku puzzle-solving.

Hopefully, this has convinced you that graph coloring is a problem worth solving. As usual, we will start by formulating an appropriate representation of a possible solution for this problem.

Solution representation

Expanding on the commonly used binary list (or array) representation, we can employ a list of integers, where each integer represents a unique color, while each element of the list matches one of the graph's nodes.

For example, since the Petersen graph has 10 nodes, we can assign each node an index between 0 and 9. Then, we can represent the node coloring for that graph using a list of 10 elements.

For example, let's have a look at what we have in this particular representation:

```
[0, 2, 1, 3, 1, 2, 0, 3, 3, 0]
```

Let's talk about what we have here in detail:

- Four colors are used, represented by the integers 0, 1, 2, and 3
- The first, seventh, and tenth nodes of the graph are colored with the first color (0)
- The third and fifth nodes are colored with the second color (1)
- The second and sixth nodes are colored with the third color (2)
- The fourth, eighth, and ninth nodes are colored with the fourth color (3)

To evaluate the solution, we need to iterate over each pair of adjacent nodes and check if they share the same color. If they do, this is a coloring violation, and we seek to minimize the number of violations to zero to achieve the proper coloring of the graph.

However, you may recall that we also wish to minimize the number of colors that are used. If we happen to already know this number, we can just use as many integer values as the known number of colors. But what if we don't? One way to go about this is to start with an estimate (or just a guess) for the number of colors used. If we find a proper solution using this number, we can reduce the number and try again. If no solution is found, we can increase the number and try again until we have the smallest number we can find a solution with. However, we may be able to get to this number faster by using soft and hard constraints, as described in the next subsection.

Using hard and soft constraints for the graph coloring problem

When solving the nurse scheduling problem earlier in this chapter, we noted the difference between hard constraints – those we have to adhere to for the solution to be considered valid – and soft constraints – those we strive to minimize to get the best solution. In the graph coloring problem, the **color assignment** requirement – where no two adjacent nodes can have the same color – is a **hard constraint**. We have to minimize the number of violations of this constraint to zero to achieve a valid solution.

Minimizing the **number of colors** used, however, can be introduced as a **soft constraint**. We would like to minimize this number, but not at the expense of violating the hard constraint.

This will allow us to launch the algorithm with several colors higher than our estimate and let the algorithm minimize it until – ideally – it reaches the actual minimal color count.

As we did for the nurse scheduling problem, we will implement this approach by creating a cost function, where the cost of a hard constraint violation is greater than the cost induced by using more colors. The total cost will then be used as the fitness function to be minimized. This functionality can be incorporated into the Python class and will be described in the next subsection.

Python problem representation

To encapsulate the graph coloring problem, we've created a Python class called GraphColoringProblem. This class can be found in the graphs.py file, which can be found at https://github.com/PacktPublishing/Hands-On-Genetic-Algorithms-with-Python-Second-Edition/blob/main/chapter_05/graphs.py.

To implement this class, we'll utilize the open source Python package **NetworkX** (https://networkx.github.io), which allows us to create, manipulate, and draw graphs, among other things. The graph we'll be using as the subject of the coloring problem is an instance of the NetworkX graph class. Instead of creating this graph from scratch, we can take advantage of the numerous preexisting graphs contained in this library, such as the *Petersen graph* we saw earlier.

The constructor of the `GraphColoringProblem` class accepts the graph to be colored as a parameter. In addition, it accepts the `hardConstraintPenalty` parameter, which represents the penalty factor for a hard constraint violation.

The constructor then creates a list of the graph's nodes, as well as an adjacency matrix, that allows us to quickly find out if any two nodes in the graph are adjacent:

```
self.nodeList = list(self.graph.nodes)
self.adjMatrix = nx.adjacency_matrix(graph).todense()
```

This class uses the following method to calculate the number of coloring violations in the given color arrangement:

- `getViolationsCount`(colorArrangement)

The following method is used to calculate the number of colors used by the given color arrangement:

- `getNumberOfColors`(colorArrangement)

In addition, the class provides the following public methods:

- `getCost`(colorArrangement): This calculates the total cost of the given color arrangement
- `plotGraph`(colorArrangement): This plots the graph with the nodes colored according to the given color arrangement

The main method of the class exercises the class' methods by creating a Petersen graph instance and testing a randomly generated color arrangement for it, containing up to five colors. In addition, it sets the value of `hardConstraintPenalty` to `10`:

```
gcp = GraphColoringProblem(nx.petersen_graph(), 10)
solution = np.random.randint(5, size=len(gcp))
```

The resulting output may look as follows:

```
solution = [2 4 1 3 0 0 2 2 0 3]
number of colors = 5
Number of violations = 1
Cost = 15
```

Since this particular random solution uses five colors and causes one coloring violation, the calculated cost is 15.

The plot for this solution is as follows – can you spot the single coloring violation?

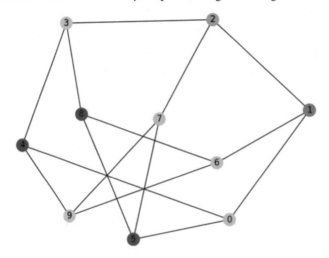

Figure 5.11: Petersen graph improperly colored with five colors

In the next subsection, we'll apply a genetic algorithm-based solution in an attempt to eliminate any coloring violations while also minimizing the number of colors that are used.

Genetic algorithms solution

To solve the graph coloring problem using a genetic algorithm, we've created a Python program called `03-solve-graphs.py`, which is located at `https://github.com/PacktPublishing/Hands-On-Genetic-Algorithms-with-Python-Second-Edition/blob/main/chapter_05/03_solve_graphs.py`.

Since the solution representation we chose for this problem is a list of integers, we need to expand the genetic approach of using a binary list.

The following steps describe the main points of our solution:

1. The program starts by creating an instance of the `GraphColoringProblem` class with the desired *NetworkX* graph to be solved – the familiar *Petersen graph* in this case – and the desired value for `hardConstraintPenalty`, which is set by the HARD_CONSTRAINT_PENALTY constant:

   ```
   gcp = graphs.GraphColoringProblem(nx.petersen_graph(),
       HARD_CONSTRAINT_PENALTY)
   ```

2. Since our goal is to minimize the cost, we'll define a single objective, minimizing fitness strategy:

   ```
   creator.create("FitnessMin", base.Fitness, weights=(-1.0,))
   ```

3. Since the solution is represented by a list of integer values representing the participating colors, we need to define a random generator that creates an integer between 0 and the number of colors minus 1. This random integer represents one of the participating colors. Then, we must define a solution (individual) creator that generates a list of these random integers that match the given graph in length – this is how we randomly assign a color for each node in the graph. Finally, we must define an operator that creates an entire population of individuals:

```
toolbox.register("Integers", random.randint, 0, MAX_COLORS - 1)
toolbox.register("individualCreator", tools.initRepeat, \
    creator.Individual, toolbox.Integers, len(gcp))
toolbox.register("populationCreator", tools.initRepeat, \
    list, toolbox.individualCreator)
```

4. The fitness evaluation function is set to calculate the combined cost of the coloring violations and the number of colors used, which is associated with each solution, by calling the getCost() method of the GraphColoringProblem class:

```
def getCost(individual):
    return gcp.getCost(individual),
toolbox.register("evaluate", getCost)
```

5. As for the genetic operators, we can still use the same *selection* and *crossover* operations we used for binary lists; however, the mutation operation needs to change. The *flip-bit mutation* that's used for binary lists flips between values of 0 and 1, while here, we need to change a given integer to another – randomly generated – integer in the allowed range. The mutUniformInt operator does just that – we just need to set the range similar to what we did with the preceding integers operator:

```
toolbox.register("select", tools.selTournament, tournsize=2)
toolbox.register("mate", tools.cxTwoPoint)
toolbox.register("mutate", tools.mutUniformInt, low=0, \
    up=MAX_COLORS - 1, indpb=1.0/len(gcp))
```

6. We keep using the *elitist approach*, where the HOF members – the current best individuals – are always passed untouched to the next generation:

```
population, logbook = elitism.eaSimpleWithElitism(\
    population, toolbox, cxpb=P_CROSSOVER, \
    mutpb=P_MUTATION, ngen=MAX_GENERATIONS, \
    stats=stats, halloffame=hof, verbose=True)
```

7. When the algorithm concludes, we print the details of the best solution that was found before plotting the graphs.

```
gcp.plotGraph(best)
```

Before we run the program, let's set the algorithm constants, as follows:

```
POPULATION_SIZE = 100
P_CROSSOVER = 0.9
P_MUTATION = 0.1
MAX_GENERATIONS = 100
HALL_OF_FAME_SIZE = 5
```

In addition, we need to set a penalty for violating hard constraints to a value of 10 and the number of colors to 10:

```
HARD_CONSTRAINT_PENALTY = 10
MAX_COLORS = 10
```

Running the program with these settings yields the following output:

```
-- Best Individual = [5, 0, 6, 5, 0, 6, 5, 0, 0, 6]
-- Best Fitness = 3.0
Number of colors = 3
Number of violations = 0
Cost = 3
```

This means that the algorithm was able to find a proper coloring for the graph using three colors, denoted by the integers 0, 5, and 6. As we mentioned previously, the actual integer values don't matter – it's the distinction between them that does. Three is indeed the known chromatic number of the Petersen graph.

The preceding code creates the following plot, which illustrates the solution's validity:

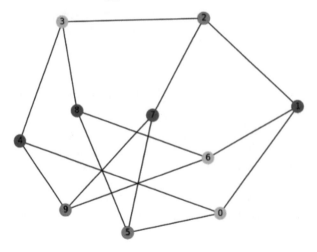

Figure 5.12: A plot of the Petersen graph properly colored by the program using three colors

The following graph, which depicts the minimum and average fitness over the generations, indicates that the algorithm reached the solution rather quickly since the Petersen graph is relatively small:

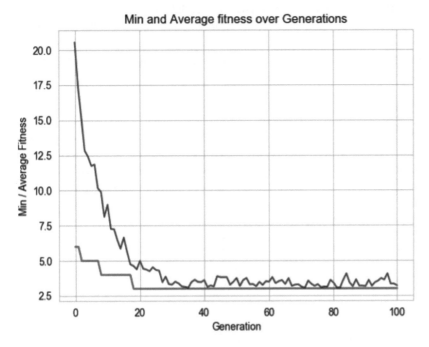

Figure 5.13: Stats of the program solving the graph coloring problem for the Petersen graph

To try a larger graph, let's replace the *Petersen graph* with a *Mycielski graph* of order 5. This graph contains 23 nodes and 71 edges, and is known to have a chromatic number of 5:

```
gcp = graphs.GraphColoringProblem(nx.mycielski_graph(5),
    HARD_CONSTRAINT_PENALTY)
```

Using the same parameters as before, including the setting of 10 colors, we get the following results:

```
-- Best Individual = [9, 6, 9, 4, 0, 0, 6, 5, 4, 5, 1, 5, 1, 1, 6, 6,
9, 5, 9, 6, 5, 1, 4]
-- Best Fitness = 6.0
Number of colors = 6
Number of violations = 0
Cost = 6
```

Since we happen to know that the chromatic number for this graph is 5, this is not the optimal solution, although it's close. How can we get there? And what if we didn't know the chromatic number beforehand? One way to go about this is to change the parameters of the genetic algorithm – for example, we could increase the population size (and possibly the HOF size) and/or increase the number of generations. Another approach would be to start the same search again but with a reduced number of colors. Since the algorithm found a solution with six colors, let's reduce the maximum number of colors to 5 and see if the algorithm can still find a valid solution:

```
MAX_COLORS = 5
```

Why would the algorithm find a five-color solution now if it didn't find one in the first place? As we lower the number of colors from 10 to 5, the search space is considerably reduced – in this case, from 10^{23} to 5^{23} (since we have 23 nodes in the graph) – and the algorithm has a better chance of finding the optimal solution(s), even with a short run and a limited population size. So, while the first run of the algorithm may get us close to the solution, it could be good practice to keep decreasing the number of colors until the algorithm can't find a better solution.

In our case, when started with five colors, the algorithm was able to find a five-color solution rather easily:

```
-- Best Individual = [0, 3, 0, 2, 4, 4, 2, 2, 2, 4, 1, 4, 3, 1, 3, 3,
4, 4, 2, 2, 4, 3, 0]
-- Best Fitness = 5.0
Number of colors = 5
Number of violations = 0
Cost = 5
```

The plot of the colored graph looks as follows:

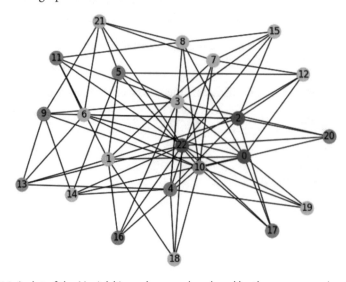

Figure 5.14: A plot of the Mycielski graph properly colored by the program using five colors

Now, if we attempt to decrease the maximum number of colors to four, we will always get at least one violation.

You are encouraged to try out other graphs and experiment with the various settings of the algorithm.

Summary

In this chapter, you were introduced to constraint satisfaction problems, which are close relatives of the previously studied combinatorial optimization problems. Then, we explored three classic constraint satisfaction cases – the *N-Queen* problem, the *nurse scheduling* problem, and the *graph coloring* problem. For each of these problems, we followed the now-familiar process of finding an appropriate representation for a solution, creating a class that encapsulates the problem and evaluates a given solution, and creating a genetic algorithm solution that utilizes that class. We ended up with valid solutions for the problems while getting acquainted with the concept of **hard constraints** versus **soft constraints**.

So far, we have been looking into discrete search problems consisting of states and state transitions. In the next chapter, we will study search problems in a continuous space, to demonstrate the versatility of the genetic algorithms approach.

Further reading

For more information on the topics that were covered in this chapter, please refer to the following resources:

- *Constraint Satisfaction Problems*, from the book *Artificial Intelligence with Python*, by Prateek Joshi, January 2017

- *Introduction to graph theory*, from the book *Python Data Science Essentials – Second Edition*, by Alberto Boschetti, Luca Massaron, October 2016

- NetworkX tutorial: `https://networkx.github.io/documentation/stable/tutorial.html`

6

Optimizing Continuous Functions

This chapter describes how **continuous search-space** optimization problems can be solved by genetic algorithms. We will start by describing the chromosomes and genetic operators commonly used for genetic algorithms with real number-based populations and go over the tools offered by the **Distributed Evolutionary Algorithms in Python** (**DEAP**) framework for this domain. We will then cover several hands-on examples of continuous function optimization problems and their Python-based solutions using the DEAP framework. These include the optimization of the *Eggholder function*, *Himmelblau's function*, as well as the constrained optimization of *Simionescu's function*. Along the way, we will learn about finding multiple solutions using **niching** and **sharing** and handling **constraints**.

By the end of this chapter, you will be able to do the following:

- Understand chromosomes and genetic operators used for real numbers
- Use DEAP to optimize continuous functions
- Optimize the Eggholder function
- Optimize Himmelblau's function
- Perform constrained optimization with Simionescu's function

Technical requirements

In this chapter, we will be using Python 3 with the following supporting libraries:

- `deap`
- `numpy`
- `matplotlib`
- `seaborn`

> **Important note**
>
> If you use the `requirements.txt` file we provide (see *Chapter 3*), these libraries are already included in your environment.

The programs used in this chapter can be found in the book's GitHub repository at the following link:

`https://github.com/PacktPublishing/Hands-On-Genetic-Algorithms-with-Python-Second-Edition/tree/main/chapter_06`

Check out the following video to see the code in action:

`https://packt.link/OEBOd`

Chromosomes and genetic operators for real numbers

In previous chapters, we focused on search problems that inherently deal with the methodic evaluation of states and transitions between states. Consequently, the solutions for these problems were best represented by lists (or arrays) of binary or integer parameters. In contrast to that, this chapter covers problems where the solution space is **continuous**, meaning the solutions are made up of real (floating-point) numbers. As we mentioned in *Chapter 2, Understanding the Key Components of Genetic Algorithms*, representing real numbers using binary or integer lists was found to be far from ideal and, instead, lists (or arrays) of real-valued numbers are now considered to be a simpler and better approach.

Reiterating the example from *Chapter 2*, if we have a problem involving three real-valued parameters, the chromosome will look like the following:

$[x_1, x_2, x_3]$

Here, x_1, x_2, x_3 represent real numbers, such as the following:

[1.23, 7.2134, -25.309] or [-30.10, 100.2, 42.424]

In addition, we mentioned that while the various *selection* methods work the same for either integer-based or real-based chromosomes, specialized *crossover* and *mutation* methods are needed for the real-coded chromosomes. These operators are usually applied on a dimension-by-dimension basis, illustrated as follows.

Suppose we have two parent chromosomes: $parent_x = [x_1, x_2, x_3]$ and $parent_y = [y_1, y_2, y_3]$. As the crossover operation is applied separately to each dimension, an offspring $[o_1, o_2, o_3]$ will be created, as follows:

- o_1 is the result of a crossover operator between x_1 and y_1
- o_2 is the result of a crossover operator between x_2 and y_2
- o_3 is the result of a crossover operator between x_3 and y_3

Similarly, the *mutation* operator will be individually applied to each dimension so that each of the components o_1, o_2, and o_3 can be subject to mutation.

Some commonly used real-coded operators are the following:

- **Blend Crossover** (also known as **BLX**), where each offspring is randomly selected from the following interval created by its parents:

$$\left[parent_x - \alpha\left(parent_y - parent_x\right), parent_y + \alpha\left(parent_y - parent_x\right)\right]$$

The α value is commonly set to 0.5, resulting in a selection interval twice as wide as the interval between the parents.

- **Simulated Binary Crossover** (**SBX**), where two offspring are created from two parents using the following formula, guaranteeing that the average of the offspring values is equal to that of the parents' values:

$$offspring_1 = \tfrac{1}{2}\left[(1 + \beta)parent_x + (1 - \beta)parent_y\right]$$
$$offspring_2 = \tfrac{1}{2}\left[(1 - \beta)parent_x + (1 + \beta)parent_y\right]$$

The value of β, also known as the *spread factor*, is calculated using a combination of a randomly chosen value and a pre-determined parameter known as η (eta), *distribution index*, or *crowding factor*. With larger values of η, offspring will tend to be more similar to their parents. Common values of η are between 10 and 20.

- **Normally distributed** (or **Gaussian**) **mutation**, where the original value is replaced with a random number that is generated using a normal distribution, with predetermined values for mean and standard deviation.

In the next section, we will see how real-coded chromosomes and genetic operators are supported by the DEAP framework.

Using DEAP with continuous functions

The DEAP framework can be used for optimizing continuous functions in a very similar manner to what we have seen so far, when we solved discrete search problems. All that's needed are a few subtle modifications.

For the chromosome encoding, we can use a list (or array) of floating-point numbers. One thing to keep in mind, though, is that the existing genetic operators of DEAP will **not** work well with individual objects extending the `numpy.ndarray` class due to the way these objects are being sliced, as well as the way they are being compared to each other.

Using `numpy.ndarray`-based individuals will require redefining the genetic operators accordingly. This is further covered in the DEAP documentation, under *Inheriting from NumPy*. For this reason, as well as for performance reasons, **ordinary** Python lists or arrays of floating-point numbers are generally **preferred** when using DEAP.

As for real-coded genetic operators, the DEAP framework offers several implementations out of the box, contained in the crossover and the mutation modules:

- `cxBlend()` is DEAP's implementation of *Blend Crossover*, using the `alpha` argument as the α value
- `cxSimulatedBinary()` implements *Simulated Binary Crossover*, using the `eta` argument as the η (crowding factor) value
- `mutGaussian()` implements *normally distributed mutation*, using the `mu` and `sigma` arguments as the values for the mean and standard deviation, respectively

In addition, since the optimization of continuous functions is typically done on a particular **bounded region** rather than on the entire space, DEAP provides a couple of operators that accept boundary parameters and guarantee that the resulting individuals will reside within these boundaries:

- `cxSimulatedBinaryBounded()` is a bounded version of the `cxSimulatedBinary()` operator, accepting the `low` and `up` arguments as the lower and upper boundaries of the search space, respectively.
- `mutPolynomialBounded()` is a bounded *mutation* operator that uses a polynomial function (instead of Gaussian) for the probability distribution. This operator also accepts the `low` and `up` arguments as the lower and upper boundaries of the search space. In addition, it uses the `eta` parameter as a crowding factor, where a high value will yield a mutant close to its original value, while a small value will produce a mutant very different from its original value.

In the next section, we will demonstrate the usage of bounded operators when optimizing a classic benchmark function.

Optimizing the Eggholder function

The Eggholder function, depicted in the following diagram, is often used as a benchmark for function optimization algorithms. Finding the single **global minimum** of this function is considered a difficult task due to the large number of local minima, which give it the eggholder shape:

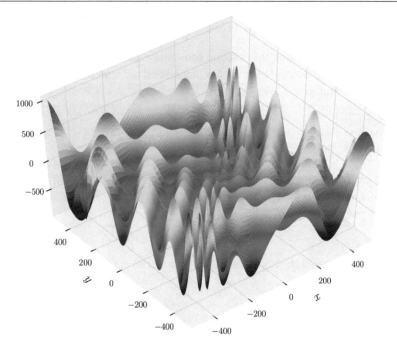

Figure 6.1: The Eggholder function

Source: https://en.wikipedia.org/wiki/File:Eggholder_function.pdf

The function can be mathematically expressed as follows:

$$f(x,y) = -(y + 47) \cdot sin\sqrt{\left|\frac{x}{2} + (y + 47)\right|} - x \cdot sin\sqrt{|x - (y + 47)|}$$

It is usually evaluated on the search space bounded by [-512, 512] in each dimension. The global minimum of the function is known to be at *x=512, y = 404.2319*, where the function's value is *-959.6407*.

In the next subsection, we will attempt to find the global minimum using the genetic algorithms method.

Optimizing the Eggholder function with genetic algorithms

The genetic algorithm-based program we created for optimizing the Eggholder function resides in the 01_optimize_eggholder.py Python program located at the following link:

https://github.com/PacktPublishing/Hands-On-Genetic-Algorithms-with-Python-Second-Edition/blob/main/chapter_06/01_optimize_eggholder.py

The following steps highlight the main parts of this program:

1. The program starts by setting the function constants, namely the number of input dimensions (2, as this function is defined over the *x-y* plane), and the search space boundaries that were mentioned previously:

```
DIMENSIONS = 2  # number of dimensions
# boundaries, same for all dimensions
BOUND_LOW, BOUND_UP = -512.0, 512.0
```

2. Since we are dealing with floating-point numbers confined by certain boundaries, we next define a helper function that creates random floating-point numbers, uniformly distributed within the given range:

> **Note**
> This function assumes that the upper and lower boundaries are the same for all dimensions.

```
def randomFloat(low, up):
    return [random.uniform(l, u) for l, \
        u in zip([low] * DIMENSIONS, [up] * DIMENSIONS)]
```

3. We next define the `attrFloat` operator. This operator utilizes the previous helper function to create a single, random floating-point number within the given boundaries. The `attrFloat` operator is then used by the `individualCreator` operator to create random individuals. This is followed by `populationCreator`, which can generate the desired number of individuals:

```
toolbox.register("attrFloat", randomFloat, BOUND_LOW, BOUND_UP)
toolbox.register("individualCreator", tools.initIterate, \
    creator.Individual, toolbox.attrFloat)
toolbox.register("populationCreator", tools.initRepeat, \
    list, toolbox.individualCreator)
```

4. Given that the object to be minimized is the Eggholder function, we use it directly as the fitness evaluator. As the individual is a list of floating-point numbers with a dimension (or length) of 2, we extract the x and y values from the individual accordingly, and then calculate the function:

```
def eggholder(individual):
    x = individual[0]
    y = individual[1]
    f = (
        -(y + 47.0) * np.sin(np.sqrt(abs(x / 2.0 + (y + 47.0))))
        - x * np.sin(np.sqrt(abs(x - (y + 47.0))))
    )
    return f,   # return a tuple
toolbox.register("evaluate", eggholder)
```

5. Next are the genetic operators. Given that the *selection* operator is independent of the individual type, and we've had a good experience so far using the *tournament selection* with a tournament size of 2, coupled with the *elitist approach*, we'll continue to use it here. The *crossover* and *mutation* operators, on the other hand, need to be specialized for floating-point numbers within given boundaries, and therefore we use the DEAP-provided `cxSimulatedBinaryBounded` operator for crossover and the `mutPolynomialBounded` operator for mutation:

```
# Genetic operators:
toolbox.register("select", tools.selTournament, tournsize=2)
toolbox.register("mate", tools.cxSimulatedBinaryBounded, \
    low=BOUND_LOW, up=BOUND_UP, eta=CROWDING_FACTOR)
toolbox.register("mutate", tools.mutPolynomialBounded, \
    low=BOUND_LOW, up=BOUND_UP, eta=CROWDING_FACTOR, \
    indpb=1.0/DIMENSIONS)
```

6. As we have done multiple times, we use our modified version of DEAP's simple genetic algorithm flow, where we added *elitism*—keeping the best individuals (members of the hall of fame) and moving them to the next generation, untouched by the genetic operators:

```
population, logbook = elitism.eaSimpleWithElitism(population,
    toolbox,
    cxpb=P_CROSSOVER,
    mutpb=P_MUTATION,
    ngen=MAX_GENERATIONS,
    stats=stats,
    halloffame=hof,
    verbose=True)
```

7. We will start with the following parameters for the genetic algorithm settings. As the Eggholder function may be somewhat difficult to optimize, we use a relatively large population size considering the low dimension count:

```
# Genetic Algorithm constants:
POPULATION_SIZE = 300
P_CROSSOVER = 0.9
P_MUTATION = 0.1
MAX_GENERATIONS = 300
HALL_OF_FAME_SIZE = 30
```

8. In addition to the previous ordinary genetic algorithm constants, we now need a new one, the **crowding factor** (eta) that is used by both the crossover and mutation operations:

```
CROWDING_FACTOR = 20.0
```

> **Important note**
>
> It is also possible to define separate crowding factors for crossover and mutation.

We are finally ready to run the program. The results obtained with these settings are shown as follows:

```
-- Best Individual = [512.0, 404.23180541839946]
-- Best Fitness = -959.6406627208509
```

This means that we have found the global minimum.

If we examine the statistics plot generated by the program, shown next, we can tell that the algorithm found some local minima values right away and then made small incremental improvements until it eventually found the global minima:

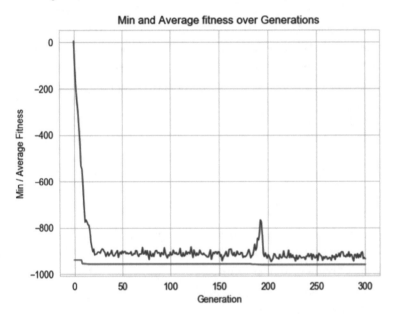

Figure 6.2: Stats of the first program optimizing the Eggholder function

One interesting area is around generation 180—let's explore it further in the next subsection.

Improving the speed with an increased mutation rate

If we zoom in at the lower part of the fitness axis, we will notice a relatively large improvement of the best result found (red line) around generation 180, accompanied by a large swing of the average results (green line):

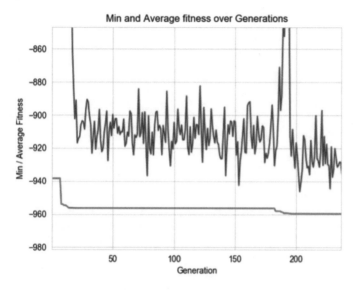

Figure 6.3: Enlarged section of the first program's stats graph

One way to interpret this observation is that perhaps introducing more noise can lead to better results faster. This could be another manifestation of the familiar principle of **exploration versus exploitation** we've discussed several times before—increasing the exploration (which manifests itself as noise in the diagram) may help us locate the global minimum faster. An easy way to increase the measure of exploration is to boost the probability of mutations. Hopefully, the use of elitism—keeping the best results untouched—will keep us from over-exploring, which leads to random search-like behavior.

To test this idea, let's increase the probability of mutation from 0.1 to 0.5:

```
P_MUTATION = 0.5
```

Running the modified program, we again found the global minimum, but much faster, as is evident from the output, as well as from the statistic plot shown next, where the red line (the best result) reaches the optimum quickly, while the average score (green) is noisier than before and is more distanced from the best result:

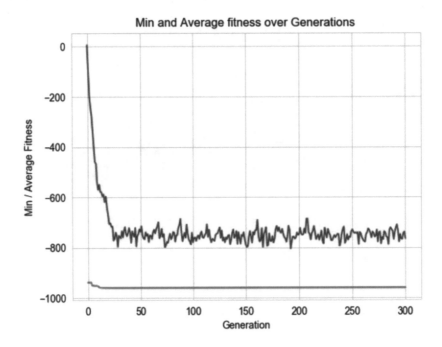

Figure 6.4: Stats of the program optimizing the Eggholder function with an increased mutation probability

We will keep this idea in mind when dealing with our next benchmark function, known as Himmelblau's function.

Optimizing Himmelblau's function

Another frequently used function for benchmarking optimization algorithms is Himmelblau's function, depicted in the following diagram:

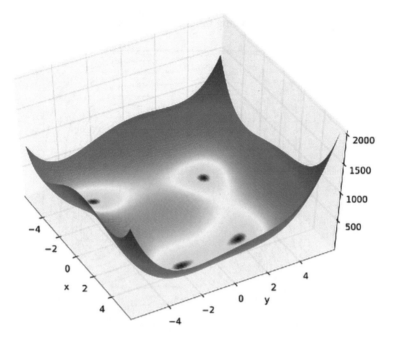

Figure 6.5: Himmelblau's function

Source: https://commons.wikimedia.org/wiki/File:Himmelblau_function.svg

Image by Morn the Gorn

The function can be mathematically expressed as follows:

$$f(x,y) = (x^2 + y - 11)^2 + (x + y^2 - 7)^2$$

It is usually evaluated on the search space bounded by [-5, 5] in each dimension.

Although this function seems simpler in comparison to the Eggholder function, it draws interest as it is **multi-modal**; in other words, it has more than one global minimum. To be exact, the function has four global minima evaluating to 0, which can be found in the following locations:

- x=3.0, y=2.0

- x=−2.805118, y=3.131312

- x=−3.779310, y=−3.283186

- x=3.584458, y=−1.848126

These locations are depicted in the following function contour diagram:

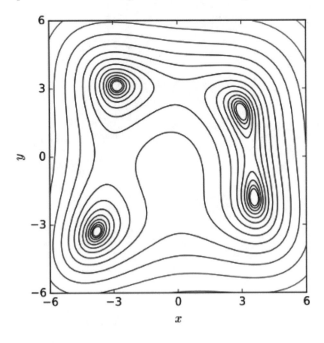

Figure 6.6: Contour diagram of Himmelblau's function

Source: `https://commons.wikimedia.org/wiki/File:Himmelblau_contour.svg`

Image by Nicoguaro

When optimizing multi-modal functions, we are often interested in finding all (or most) minima locations. However, let's start with finding one, which we are going to do in the next subsection.

Optimizing Himmelblau's function with genetic algorithms

The genetic algorithm-based program we created for finding a single minimum of Himmelblau's function resides in the `02_optimize_himmelblau.py` Python program, located at the following link:

`https://github.com/PacktPublishing/Hands-On-Genetic-Algorithms-with-Python-Second-Edition/blob/main/chapter_06/02_optimize_himmelblau.py`

The program is similar to the one we used for optimizing the Eggholder function, with a few differences highlighted as follows:

1. We set the boundaries for this function to [-5.0, 5.0]:

    ```
    BOUND_LOW, BOUND_UP = -5.0, 5.0  # boundaries for all dimensions
    ```

2. We now use Himmelblau's function as the fitness evaluator:

```
def himmelblau(individual):
    x = individual[0]
    y = individual[1]
    f = (x ** 2 + y - 11) ** 2 + (x + y ** 2 - 7) ** 2
    return f,  # return a tuple
toolbox.register("evaluate", himmelblau)
```

3. Since the function we optimize has several minima, it may be interesting to observe the distribution of the solutions found at the end of the run. We, therefore, add a scatter graph containing the locations of the four global minima and the final population on the same *x-y* plane:

```
plt.figure(1)
globalMinima = [[3.0, 2.0], [-2.805118, 3.131312],
        [-3.779310, -3.283186], [3.584458, -1.848126]]
plt.scatter(*zip(*globalMinima), marker='X', color='red',
    zorder=1)
plt.scatter(*zip(*population), marker='.', color='blue',
    zorder=0)
```

4. We also print the members of the hall of fame—the best individuals found during the run:

```
print("- Best solutions are:")
for i in range(HALL_OF_FAME_SIZE):
    print(i, ": ", hof.items[i].fitness.values[0],
            " -> ", hof.items[i])
```

Running the program, the results indicate that we found one of the four minima (x=3.0, y=2.0):

```
-- Best Individual = [2.999999999987943, 2.0000000000007114]
-- Best Fitness = 4.523490304795033e-23
```

The printout of the hall-of-fame members suggests they all represent the same solution:

```
- Best solutions are:
0 : 4.523490304795033e-23 -> [2.999999999987943, 2.0000000000007114]
1 : 4.523732642865117e-23 -> [2.999999999987943, 2.000000000000697]
2 : 4.523900512465748e-23 -> [2.999999999987943, 2.0000000000006937]
3 : 4.524063333565856e-23 -> [2.999999999987943, 2.00000000000071]
...
```

The following diagram, illustrating the distribution of the entire population, further confirms that the genetic algorithms have converged to one of the four functions' minima—the one residing at (x=3.0, y=2.0):

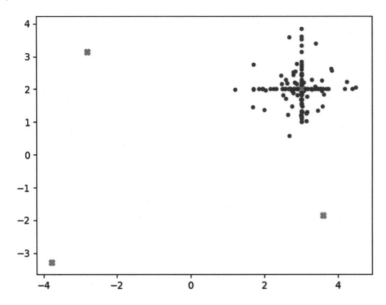

Figure 6.7: Scatter graph of the population at the end of the first run, alongside the four functions' minima

In addition, it is evident that many of the individuals in the population have either the x or the y component of the minima we found.

These results represent what we generally expect from the genetic algorithm—to identify a global optimum and converge to it. Since, in this case, we have several minima, it is expected to converge to one of them. Which one it will be is largely based on the random initialization of the algorithm. As you may recall, in all our programs so far, we have been using a fixed random seed (of value 42):

```
RANDOM_SEED = 42
random.seed(RANDOM_SEED)
```

This is done to enable the repeatability of the results; however, in real life, we will typically use different random seed values for different runs, either by commenting out these lines or by explicitly setting the constant to different values.

For example, if we set the seed value to 13, we will end up with the solution (x=−2.805118, y=3.131312), as illustrated in the following diagram:

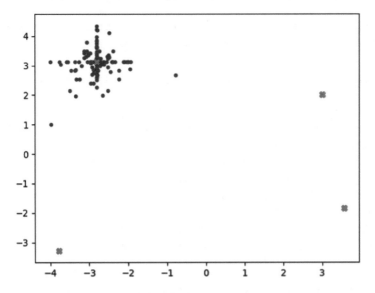

Figure 6.8: Scatter graph of the population at the end of the
second run, alongside the four functions' minima

If we proceed to change the seed value to 17, the program execution will yield the solution (x=3.584458, y=−1.848126), as illustrated by the following diagram:

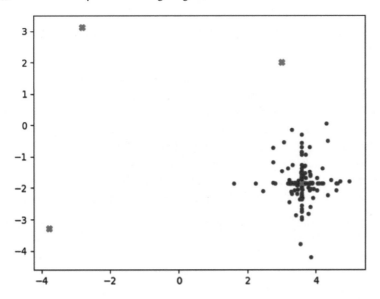

Figure 6.9: Scatter graph of the population at the end of the third run, alongside the four functions' minima

However, what if we wanted to find *all* global minima in a single run? As we will see in the next subsection, genetic algorithms offer us a way to pursue this goal.

Using niching and sharing to find multiple solutions

In *Chapter 2, Understanding the Key Components of Genetic Algorithms*, we mentioned that **niching** and **sharing** in genetic algorithms mimic the way a natural environment is divided into multiple sub-environments or *niches*. These niches are populated by different species, or sub-populations, taking advantage of the unique resources available in each niche, while specimens that coexist in the same niche have to compete over the same resources. Implementing a sharing mechanism within the genetic algorithm will encourage individuals to explore new niches and can be used for finding several optimal solutions, each considered a niche. One common way to accomplish sharing is to divide the raw fitness value of each individual with (some function of) the combined distances from all the other individuals, effectively penalizing a crowded population by sharing the local bounty between its individuals.

Let's try to apply this idea to Himmelblau's function optimization process and see whether it can help locate all four minima in a single run. This attempt is implemented in the `03_optimize_himmelblau_sharing.py` program, located at the following link:

`https://github.com/PacktPublishing/Hands-On-Genetic-Algorithms-with-Python-Second-Edition/blob/main/chapter_06/03_optimize_himmelblau_sharing.py`

The program is based on the previous one, but we had to make some important modifications, described as follows:

1. For starters, the implementation of a sharing mechanism usually requires us to optimize a function that produces positive fitness values and to look for *maxima* values rather than *minima*. This enables us to divide the raw fitness values as a way to decrease fitness and practically share the resources between neighboring individuals. As Himmelblau's function produces values between 0 and (roughly) 2,000, we can instead use a modified function that returns 2,000 minus the original value, which will guarantee that all function values are positive, while transforming the minima points into maxima points that return the value of 2,000. As the locations of these points are not going to change, finding them will still serve our original purpose:

```
def himmelblauInverted(individual):
    x = individual[0]
    y = individual[1]
    f = (x ** 2 + y - 11) ** 2 + (x + y ** 2 - 7) ** 2
    return 2000.0 - f,  # return a tuple
toolbox.register("evaluate", himmelblauInverted)
```

2. To complete the conversion, we redefine the fitness strategy to be a *maximizing* one:

```
creator.create("FitnessMax", base.Fitness, weights=(1.0,))
```

3. To enable the implementation of *sharing*, we first create two additional constants:

```
DISTANCE_THRESHOLD = 0.1
SHARING_EXTENT = 5.0
```

4. Next, we need to implement the sharing mechanism. One convenient location for this implementation is within the *selection* genetic operator. The selection operator is where the fitness values of all individuals are examined and used to select the parents for the next generation. This enables us to inject some code that recalculates these fitness values just before the selection takes place and then retrieves the original fitness values before continuing, for the purpose of tracking. To make this happen, we implemented a new `selTournamentWithSharing()` function, which has the same signature as the original `tools.selTournament()` function we have been using until now:

```
def selTournamentWithSharing(individuals, k, tournsize,
    fit_attr="fitness"):
```

This function starts by setting the original fitnesses aside so that they can be retrieved later. It then iterates over each individual and calculates a number, `sharingSum`, by which its fitness value will be divided. This sum value is accumulated by calculating the distance between the location of the current individual and the location of each of the other individuals in the population. If the distance is smaller than the threshold defined by the `DISTANCE_THRESHOLD` constant, the following value is added to the accumulating sum:

$$1 - \frac{distance}{DISTANCE\_THRESHOLD} \times \frac{1}{SHARING\_EXTENT}$$

This means that the *reduction* in the fitness value will be greater in the following scenarios:

- The (normalized) distance between the individuals is smaller

- The value of the `SHARING_EXTENT` constant is larger

After recalculating the fitness value for each individual, *tournament selection* is conducted using the new fitness values:

```
selected = tools.selTournament(individuals, k, tournsize,
    fit_attr)
```

Lastly, the original fitness values are retrieved:

```
for i, ind in enumerate(individuals):
    ind.fitness.values = origFitnesses[i],
```

5. As a final touch, we added a plot showing the locations of the best individuals—the hall-of-fame members—on the *x-y* plane, alongside the known optima location, similar to what we already do for the entire population:

```
plt.figure(2)
plt.scatter(*zip(*globalMaxima), marker='x', color='red',
    zorder=1)
plt.scatter(*zip(*hof.items), marker='.', color='blue',
    zorder=0)
```

When we run this program, the results don't disappoint. Examining the members of the hall of fame, it seems that we have located all four optima locations:

```
 - Best solutions are:
 0 : 1999.9997428476076 -> [3.00161237138945, 1.9958270919300878]
 1 : 1999.9995532774788 -> [3.585506608049694, -1.8432407550446581]
 2 : 1999.9988186889173 -> [3.585506608049694, -1.8396197402430106]
 3 : 1999.9987642838498 -> [-3.7758887140006174, -3.285804345540637]
 4 : 1999.9986563457114 -> [-2.8072634380293766, 3.125893564009283]
 ...
```

The following diagram, illustrating the distribution of the hall-of-fame members, further confirms that:

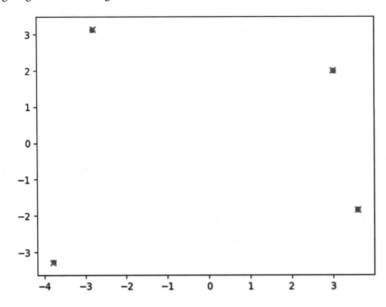

Figure 6.10: Scatter graph of the best solutions at the end of the run,
alongside the four functions' minima, when using niching

Meanwhile, the diagram depicting the distribution of the *entire* population demonstrates how the population is scattered around the four solutions:

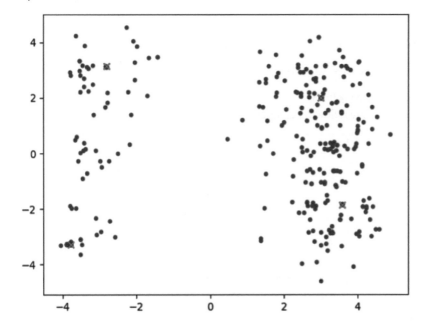

Figure 6.11: Scatter graph of the population at the end of the run,
alongside the four functions' minima, when using niching

As impressive as this may seem, we need to remember that what we did here can prove harder to implement in real-life situations. For one, the modifications we added to the selection process increase the calculation complexity and the time consumed by the algorithm. In addition, the population size usually needs to be increased so that it can sufficiently cover all areas of interest. The values of the sharing constants may be difficult to determine in some cases—for example, if we don't know in advance how close together the various peaks may be. However, we can always use this technique to roughly locate areas of interest and then further explore each one of them using the standard version of the algorithm.

An alternative approach for finding several optima points falls within the realm of **constrained optimization**, which is the subject of the next section.

Simionescu's function and constrained optimization

At first glance, Simionescu's function may not look particularly interesting. However, it has a constraint attached to it that makes it intriguing to work with as well as pleasant to look at.

The function is usually evaluated on the search space bounded by [-1.25, 1.25] in each dimension and can be mathematically expressed as follows:

$$f(x,y) = 0.1xy$$

Here, the values of x, y are subject to the following condition:

$$x^2 + y^2 \leq \left[1 + 0.2 \cdot \cos\left(8 \cdot \arctan\frac{x}{y}\right)\right]^2$$

This constraint effectively limits the values of x and y that are considered valid for this function. The result is depicted in the following contour diagram:

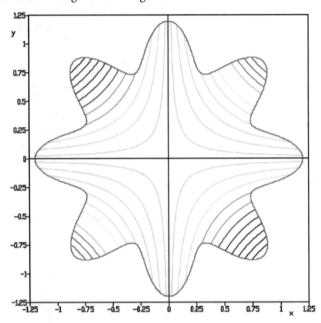

Figure 6.12: Contour diagram of the constrained Simionescu's function

Source: https://commons.wikimedia.org/wiki/File:Simionescu%27s_function.PNG

Image by Simiprof

The flower-shaped border is created by the constraint, while the colors of the contours denote the actual value—red for the highest values and purple for the lowest. If it weren't for the constraint, the minima points would have been at (1.25, -1.25) and (-1.25, 1.25). However, after applying the constraint, the global minima of the function are located at the following locations:

- $x=0.84852813$, $y=-0.84852813$
- $x=-0.84852813$, $y=0.84852813$

These represent the tips of the two opposite petals containing the purple contours. Both minima evaluate to the value of -0.072.

In the next subsection, we will attempt to find these minima using our real-coded genetic algorithms approach.

Constrained optimization with genetic algorithms

We have already dealt with constraints in *Chapter 5, Constraint Satisfaction*, when we tackled constraints within the realm of search problems. However, while search problems presented us with invalid states or combinations, here we need to address constraints in the continuous space, defined as mathematical inequalities.

The approaches for both cases, however, are similar, and the differences lie in the implementation. Let's revisit these approaches:

- The best approach, when available, is to eliminate the possibility of a constraint violation. We have actually been doing it all along in this chapter as we have used bounded regions for our functions. These are actually simple constraints on each input variable. We were able to go around them by generating initial populations within the given boundaries and by utilizing bounded genetic operators such as `cxSimulatedBinaryBounded()`, which produced results within the given boundaries. Unfortunately, this approach can prove difficult to implement when the constraints are more complex than just the upper and lower bounds for an input variable.

- Another approach is to discard candidate solutions that violate any given constraint. As we mentioned before, this approach leads to the loss of information contained in these solutions and can considerably slow down the optimization process.

- The next approach is to repair any candidate solution that violates a constraint by modifying it so it will no longer violate the constraint(s). This can prove difficult to implement and, at the same time, may lead to significant loss of information.

- Finally, the approach that worked for us in *Chapter 5, Constraint Satisfaction*, was to penalize candidate solutions that violated a constraint by degrading the solution's score and making it less desirable. For search problems, we implemented this approach by creating a cost function that added a fixed cost to each constraint violation. Here, in the continuous space case, we can either use a fixed penalty or increase the penalty based on the degree to which the constraint was violated.

When taking the last approach—penalizing the score for constraint violations—we can utilize a feature offered by the DEAP framework, namely the **penalty function**, as we will demonstrate in the next subsection.

Optimizing Simionescu's function using genetic algorithms

The genetic algorithm-based program we created for optimizing Simionescu's function resides in the `04_optimize_simionescu.py` Python program, located at the following link:

`https://github.com/PacktPublishing/Hands-On-Genetic-Algorithms-with-Python-Second-Edition/blob/main/chapter_06/04_optimize_simionescu.py`

The program is very similar to the first one we used in this chapter, created originally for the Eggholder function, with the following highlighted differences:

1. The constants setting the boundaries are adjusted to match the domain of Simionescu's function:

    ```
    BOUND_LOW, BOUND_UP = -1.25, 1.25
    ```

2. In addition, a new constant determines a fixed penalty (or cost) for violating the constraint:

    ```
    PENALTY_VALUE = 10.0
    ```

3. The fitness is now determined by the definition of Simionescu's function:

    ```
    def simionescu(individual):
        x = individual[0]
        y = individual[1]
        f = 0.1 * x * y
        return f,   # return a tuple
    toolbox.register("evaluate", simionescu)
    ```

4. Here is where the interesting part begins: we now define a new `feasible()` function that specifies the valid input domain using the constraints. This function returns a value of `True` for *x, y* values that comply with the constraints, and a value of `False` otherwise:

    ```
    def feasible(individual):
        x = individual[0]
        y = individual[1]
        return x**2 + y**2 <=
            (1 + 0.2 * math.cos(8.0 * math.atan2(x, y)))**2
    ```

5. We then use DEAP's `toolbox.decorate()` operator in combination with the `tools.DeltaPenalty()` function to modify (*decorate*) the original fitness function so that the fitness values will be penalized whenever the constraints are not satisfied. `DeltaPenalty()` accepts the `feasible()` function and the fixed penalty value as parameters:

    ```
    toolbox.decorate("evaluate", tools.DeltaPenalty(
        feasible, PENALTY_VALUE))
    ```

> **Important note**
> The `DeltaPenalty()` function can also accept a third parameter that represents the distance from the feasible region, causing the penalty to increase with the distance.

Now, the program is ready to use! The results indicate that we have indeed found one of the two known minima locations:

```
-- Best Individual = [0.8487712463169383, -0.8482833185888866]
-- Best Fitness = -0.07199984895485578
```

What about the second location? Read on—we will be looking for it in the next subsection.

Using constraints to find multiple solutions

Earlier in this chapter, when optimizing Himmelblau's function, we were looking for more than one minimum location, and observed two possible ways to do that—one was changing the random seed, and the other was using **niching and sharing**. Here, we will demonstrate a third option, powered by... constraints!

The niching technique we used for Himmelblau's function is sometimes called *parallel niching* as it attempts to locate several solutions at the same time. As we already mentioned, it is prone to several practical drawbacks. *Serial niching* (or *sequential niching*), on the other hand, is a method used to find one solution at a time. To implement serial niching, we use the genetic algorithm as usual and find the best solution. We then update the fitness function so that the area of the solution(s) already found is penalized, thereby encouraging the algorithm to explore other areas of the problem space. This can be repeated multiple times until no additional viable solutions are found.

Interestingly, penalizing the areas around the previously found solutions can be implemented by imposing constraints on the search space and, as we just learned how to apply constraints to the function at hand, we can use this knowledge to implement serial niching, demonstrated as follows.

To find the second minimum for Simionescu's function, we created the `05_optimize_simionescu_second.py` Python program, located at the following link:

```
https://github.com/PacktPublishing/Hands-On-Genetic-Algorithms-with-
Python-Second-Edition/blob/main/chapter_06/05_optimize_simionescu_
second.py
```

The program is almost identical to the previous one, with a couple of changes, as follows:

1. We first add a constant that defines the *distance threshold* from previously found solutions—new solutions that are closer than this threshold value to any of the old ones will be penalized:

    ```
    DISTANCE_THRESHOLD = 0.1
    ```

2. We then add a second constraint to the definition of the `feasible()` function using a conditional statement with multiple clauses. The new constraint applies to input values closer than the threshold to the already found solution (x=0.848, y = -0.848):

```
def feasible(individual):
    x = individual[0]
    y = individual[1]
    if x**2 + y**2 > (1 + 0.2 * math.cos(
        8.0 * math.atan2(x, y))
    )**2:
        return False
    elif (x - 0.848)**2 + (y + 0.848)**2 <
        DISTANCE_THRESHOLD**2:
        return False
    else:
        return True
```

When running this program, the results indicate that we have indeed found the second minimum:

```
-- Best Individual = [-0.8473430282562487, 0.8496942440090975]
-- Best Fitness = -0.07199824938105727
```

You are encouraged to add this minimum point as another constraint to the `feasible()` function and verify that running the program again does *not* find any other equally minimum-valued locations in the input space.

Summary

In this chapter, you were introduced to continuous search-space optimization problems and how they can be represented and solved using genetic algorithms, specifically by utilizing the DEAP framework. We then explored several hands-on examples of continuous function optimization problems—the Eggholder function, Himmelblau's function, and Simionescu's function—along with their Python-based solutions. In addition, we covered approaches for finding multiple solutions and for handling constraints.

In the next four chapters of the book, we will demonstrate how the various techniques we've learned so far in this book can be applied when solving **machine learning** (**ML**)- and **artificial intelligence** (**AI**)-related problems. The first of these chapters will provide a quick overview of **supervised learning** (**SL**) and then demonstrate how genetic algorithms can improve the outcome of learning models by selecting the most relevant portions of the given dataset.

Further reading

For more information, please refer to the following resources:

- *Mathematical optimization: finding minima of functions*:

  `http://scipy-lectures.org/advanced/mathematical_optimization/`

- *Optimization Test Functions and Datasets*:

  `https://www.sfu.ca/~ssurjano/optimization.html`

- *Introduction to Constrained Optimization*:

  `https://web.stanford.edu/group/sisl/k12/optimization/MO-unit3-pdfs/3.1introandgraphical.pdf`

- Constraint handling in DEAP:

  `https://deap.readthedocs.io/en/master/tutorials/advanced/constraints.html`

Part 3:
Artificial Intelligence
Applications of
Genetic Algorithms

This part focuses on using genetic algorithms to enhance various artificial intelligence tasks, including machine learning and natural language processing. It starts by showcasing how these algorithms can enhance supervised learning models through optimal feature selection for regression and classification tasks. We then explore the improvement of model performance through hyperparameter tuning, comparing traditional grid search methods with genetic algorithm approaches. We then shift our focus to the optimization of artificial neural network architectures, using the Iris dataset to illustrate the combined optimization of network structure and hyperparameters. In the realm of reinforcement learning, genetic algorithms are then applied to tackle Gymnasium's MountainCar and CartPole challenges, while for natural language processing we see genetic algorithms in action solving a mystery word game and assisting in document classification. Finally, we examine the use of genetic algorithms in creating "what-if" scenarios in datasets, employing counterfactual analysis in Explainable AI and causality.

This part contains the following chapters:

- *Chapter 7, Enhancing Machine Learning Models Using Feature Selection*
- *Chapter 8, Hyperparameter Tuning of Machine Learning Models*
- *Chapter 9, Architecture Optimization of Deep Learning Networks*
- *Chapter 10, Reinforcement Learning with Genetic Algorithms*
- *Chapter 11, Natural Language Processing*
- *Chapter 12, Explainable AI and Counterfactuals*

7
Enhancing Machine Learning Models Using Feature Selection

This chapter describes how genetic algorithms can be used to improve the performance of **supervised machine learning** models by selecting the best subset of features from the provided input data. We will start with a brief introduction to machine learning and then describe the two main types of supervised machine learning tasks – **regression** and **classification**. We will then discuss the potential benefits of **feature selection** when it comes to the performance of these models. Next, we will demonstrate how genetic algorithms can be utilized to pinpoint the genuine features that are generated by the **Friedman-1 Test** regression problem. Then, we will use the real-life **Zoo dataset** to create a classification model and improve its accuracy – again by applying genetic algorithms to isolate the best features for the task.

In this chapter, we will cover the following topics:

- Understand the basic concepts of supervised machine learning, as well as regression and classification tasks

- Understand the benefits of feature selection on the performance of supervised learning models

- Enhance the performance of a regression model for the Friedman-1 Test regression problem, using feature selection carried out by a genetic algorithm coded with the DEAP framework

- Enhance the performance of a classification model for the Zoo dataset classification problem, using feature selection carried out by a genetic algorithm coded with the DEAP framework

We will start this chapter with a quick review of supervised machine learning. If you are a seasoned data scientist, feel free to skip the introductory sections.

Technical requirements

In this chapter, we will be using Python 3 with the following supporting libraries:

- `deap`
- `numpy`
- `pandas`
- `matplotlib`
- `seaborn`
- `scikit-learn` – introduced in this chapter

> **Important Note**
>
> If you use the `requirements.txt` file we provide (see *Chapter 3*), these libraries are already included in your environment.

In addition, we will be using the *UCI Zoo Dataset* (`https://archive.ics.uci.edu/ml/datasets/zoo`).

The programs that will be used in this chapter can be found in this book's GitHub repository at `https://github.com/PacktPublishing/Hands-On-Genetic-Algorithms-with-Python-Second-Edition/tree/main/chapter_07`.

Check out the following video to see the code in action:

`https://packt.link/OEBOd`.

Supervised machine learning

The term **machine learning** typically refers to a computer program that receives input and produces output. Our goal is to train this program, also known as the **model**, to produce the correct output for the given input *without explicitly programming it*.

During this training process, the model learns the mapping between the inputs and the outputs by adjusting its internal parameters. One common way to train the model is by providing it with a set of inputs for which the correct output is known. For each of these inputs, we tell the model what the correct output is so that it can adjust, or tune itself, aiming to eventually produce the desired output for each of the given inputs. This tuning is at the heart of the learning process.

Over the years, many types of machine learning models have been developed. Each model has its own particular internal parameters that can affect the mapping between the input and the output, and the values of these parameters can be tuned, as illustrated in the following diagram:

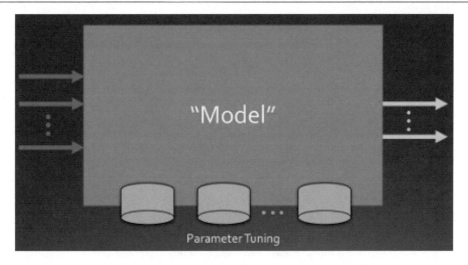

Figure 7.1: Parameter tuning of a machine learning model

For example, if the model was implementing a *decision tree*, it could contain several `IF- THEN` statements, which can be formulated as follows:

```
IF <input value> IS LESS THEN <some threshold value>
    THEN <go to some target branch>
```

In this case, both the threshold value and the identity of the target branch are parameters that can be adjusted, or tuned, during the learning process.

To tune the internal parameters, each type of model has an accompanying *learning algorithm* that iterates over the given input and output values and seeks to match the given output for each of the given inputs. To accomplish this goal, a typical learning algorithm will measure the difference (also called *error*, or more generally *loss*) between the actual output and the desired output; the algorithm will then attempt to minimize this error by adjusting the model's internal parameters.

The two main types of supervised machine learning are **classification** and **regression**, and will be described in the following subsections.

Classification

When carrying out a classification task, the model needs to decide which *category* a certain input belongs to. Each category is represented by a single output (called a **label**), while the inputs are called **features**:

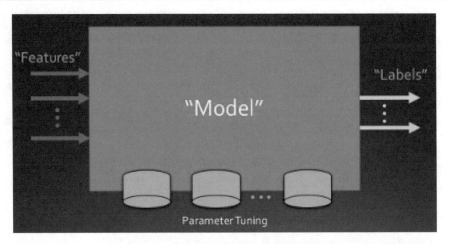

Figure 7.2: Machine learning classification model

For example, in the well-known *Iris dataset* (`https://archive.ics.uci.edu/ml/datasets/Iris`), there are four features: **Petal length**, **Petal width**, **Sepal length**, and **Sepal width**. These represent the measurements that have been manually taken of actual Iris flowers.

In terms of the output, there are three labels: **Iris setosa**, **Iris virginica**, and **Iris versicolor**. These represent the three different types of Iris in the dataset.

When input values, which represent the measurements that were taken from a given Iris flower, are present we expect the output of the correct label to go high and the other two to go low:

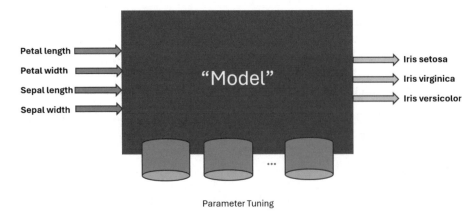

Figure 7.3: Iris Flower classifier illustrated

Classification tasks have a multitude of real-life applications, such as approval of bank loans and credit cards, email spam detection, handwritten digit recognition, and face recognition. Later in this chapter, we will be demonstrating the classification of animal types using the *Zoo dataset*.

The second main type of supervised machine learning, **regression**, will be described in the next subsection.

Regression

In contrast to classification tasks, models for regression tasks map the input values into a **single output** to provide a continuous value, as illustrated in the following diagram:

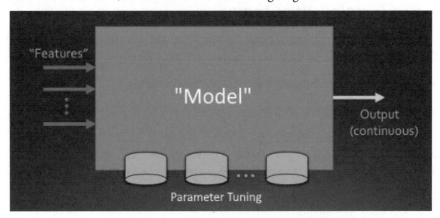

Figure 7.4: Machine learning regression model

Given the input values, the model is expected to predict the correct value of the output.

Real-life examples of regression include predicting the value of stocks, the quality of wine, or the market price of a house, as depicted in the following diagram:

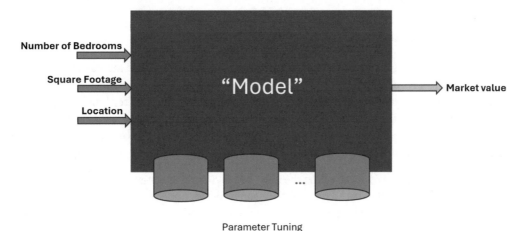

Figure 7.5: House pricing regressor

In the preceding image, the inputs are features that provide information that describes a given house, while the output is the predicted value of the house.

Many types of models exist for carrying out classification and regression tasks – some of them are described in the following subsection.

Supervised learning algorithms

As we mentioned previously, each supervised learning model consists of a set of internal tunable parameters and an algorithm that tunes these parameters in an attempt to achieve the required result.

Some common supervised learning models/algorithms are as follows:

- **Support Vector Machines (SVMs)**: Algorithms that map the given inputs as points in space so that the inputs that belong to separate categories are divided by the largest possible gap.

- **Decision Trees**: A family of algorithms that utilize a tree-like graph, where branching points represent decisions and the branches represent their consequences.

- **Random Forests**: Algorithms that create a large number of decision trees during the training phase and use a combination of their outputs.

- **Artificial Neural Networks**: Models that consist of multiple simple nodes, or neurons, which can be interconnected in various ways. Each connection can have a weight that controls the level of the signal that's carried from one neuron to the next.

There are certain techniques that can be used to improve and enhance the performance of such models. One interesting technique – **feature selection** – will be discussed in the next section.

Feature selection in supervised learning

As we saw in the previous section, a supervised learning model receives a set of inputs, called **features**, and maps them to a set of outputs. The assumption is that the information described by the features is useful for determining the value of the corresponding outputs. At first glance, it may seem that the more information we can use as input, the better our chances of predicting the output(s) correctly. However, in many cases, the opposite holds true; if some of the features we use are irrelevant or redundant, the consequence could be a (sometimes significant) decrease in the accuracy of the models.

Feature selection is the process of selecting the most beneficial and essential set of features out of the entire given set of features. Besides increasing the accuracy of the model, a successful feature selection can provide the following advantages:

- The training times of the models are shorter.

- The resulting trained models are simpler and easier to interpret.

- The resulting models are likely to provide better generalization, that is, they perform better with new input data that is dissimilar to the data that was used for training.

When looking at methods to carry out feature selection, genetic algorithms are a natural candidate. We will demonstrate how they can be applied to find the best features out of an artificially generated dataset in the next section.

Selecting the features for the Friedman-1 regression problem

The *Friedman-1* regression problem, which was created by Friedman and Breiman, describes a single output value, y, which is a function of five input values, x_0, x_1, x_2, x_3, x_4, and randomly generated noise, according to the following formula:

$$y(x_0, x_1, x_2, x_3, x_4)$$
$$= 10 \cdot \sin(\pi \cdot x_0 \cdot x_1) + 20(x_2 - 0.5)^2 + 10x_3 + 5x_4 + noise$$
$$\cdot N(0, 1)$$

The input variables, $x_0..x_4$, are independent, and uniformly distributed over the interval [0, 1]. The last component in the formula is the randomly generated noise. The noise is **normally distributed** and multiplied by the constant *noise*, which determines its level.

In Python, the `scikit-learn (sklearn)` library provides us with the `make_friedman1()` function, which can be used to generate a dataset containing the desired number of samples. Each of the samples consists of randomly generated $x_0..x_4$ values and their corresponding calculated y value. The interesting part, however, is that we can tell the function to add an arbitrary number of irrelevant input variables to the five original ones by setting the `n_features` parameter of the function to a value larger than five. If, for example, we set the value of `n_features` to 15, we will get a dataset containing the original five input variables (or features) that were used to generate the y values according to the preceding formula and an additional 10 features that are completely irrelevant to the output. This can be used, for example, to test the resilience of various regression models to noise and the presence of irrelevant features in the dataset.

We can take advantage of this function to test the effectiveness of genetic algorithms as a feature selection mechanism. In our test, we will use the `make_friedman1()` function to create a dataset with 15 features and use the genetic algorithm to search for the subset of features that provides the best performance. As a result, we expect the genetic algorithm to pick the first five features and drop the rest, assuming that the model's accuracy is better when only the relevant features are used as input. The fitness function of the genetic algorithm will utilize a regression model that, for each potential solution a subset of the original features will be trained using the dataset containing only the selected features.

As usual, we will start by choosing an appropriate representation for the solution, as described in the next subsection.

Solution representation

The objective of our algorithm is to find a subset of features that yield the best performance. Therefore, a solution needs to indicate which features are chosen and which are dropped. One obvious way to go about this is to represent each individual using a **list of binary values**. Every entry in that list corresponds to one of the features in the dataset. A value of 1 represents selecting the corresponding feature, while a value of 0 means that the feature has not been selected. This is very similar to the approach we used in the **knapsack 0-1 problem** we described in *Chapter 4, Combinatorial Optimization*.

The presence of each 0 in the solution will be translated into dropping the corresponding feature's data column from the dataset, as we will see in the next subsection.

Python problem representation

To encapsulate the Friedman-1 feature selection problem, we've created a Python class called `Friedman1Test`. This class can be found in the `friedman.py` file, which is located at `https://github.com/PacktPublishing/Hands-On-Genetic-Algorithms-with-Python-Second-Edition/blob/main/chapter_07/friedman.py`.

The main parts of this class are as follows:

1. The `__init__()` method of the class creates the dataset, as follows:

    ```
    self.X, self.y = datasets.make_friedman1(
        n_samples=self.numSamples,
        n_features=self.numFeatures,
        noise=self.NOISE,
        random_state=self.randomSeed)
    ```

2. Then, it divides the data into two subsets—a training set and a validation set—using the scikit-learn `model_selection.train_test_split()` method:

    ```
    self.X_train,self.X_validation,self.y_train,self.y_validation = \
        model_selection.train_test_split(self.X, self.y,
            test_size=self.VALIDATION_SIZE,
            random_state=self.randomSeed)
    ```

    Dividing the data into a **train set** and a **validation set** allows us to train the regression model on the train set, where the correct prediction is given to the model for training purposes, and then test it with the separate validation set, where the correct predictions are not given to the model and are, instead, compared to the predictions it produces. This way, we can test how well the model is able to generalize, rather than memorize, the training data.

3. Next, we create the regression model, for which we chose the **Gradient Boosting Regressor (GBR)** type. This model creates an **ensemble** (or aggregation) of decision trees during the training phase:

```
self.regressor = GradientBoostingRegressor(\
    random_state=self.randomSeed)
```

> **Important Note**
>
> In our example, we are passing the random seed along so that it can be used internally by the regressor. This way, we can make sure the results that we obtain are repeatable.

4. The `getMSE()` method of the class is used to determine the performance of our gradient-boosting regression model for a set of selected features. It accepts a list of binary values corresponding to the features in the dataset—a value of 1 represents selecting the corresponding feature, while a value of 0 means that the feature is dropped. The method then deletes the columns in the training and validation sets that correspond to the unselected features:

```
zeroIndices = [i for i, n in enumerate(zeroOneList) if n == 0]
currentX_train = np.delete(self.X_train, zeroIndices, 1)
currentX_validation = np.delete(self.X_validation,
    zeroIndices, 1)
```

5. The modified train set—containing only the selected features—is then used to train the regressor, while the modified validation set is used to evaluate its predictions:

```
self.regressor.fit(currentX_train, self.y_train)
prediction = self.regressor.predict(currentX_validation)
return mean_squared_error(self.y_validation, prediction)
```

The metric used here to evaluate the regressor is called the **mean square error** (MSE), which finds the average squared difference between the model's predicted values and the actual values. A *lower* value of this measurement indicates *better* performance of the regressor.

6. The `main()` method of the class creates an instance of the `Friedman1Test` class with 15 features. Then, it repeatedly uses the `getMSE()` method to evaluate the performance of the regressor with the first n features, while n is incremented from 1 to 15:

```
for n in range(1, len(test) + 1):
    nFirstFeatures = [1] * n + [0] * (len(test) - n)
    score = test.getMSE(nFirstFeatures)
```

When running the main method, the results show that, as we add the first five features one by one, the performance improves. However, afterward, each additional feature degrades the performance of the regressor:

```
 1 first features:  score = 47.553993
 2 first features:  score = 26.121143
 3 first features:  score = 18.509415
 4 first features:  score = 7.322589
 5 first features:  score = 6.702669
 6 first features:  score = 7.677197
 7 first features:  score = 11.614536
 8 first features:  score = 11.294010
 9 first features:  score = 10.858028
10 first features:  score = 11.602919
11 first features:  score = 15.017591
12 first features:  score = 14.258221
13 first features:  score = 15.274851
14 first features:  score = 15.726690
15 first features:  score = 17.187479
```

This is further illustrated by the generated plot, showing the minimum MSE value where the first five features are used:

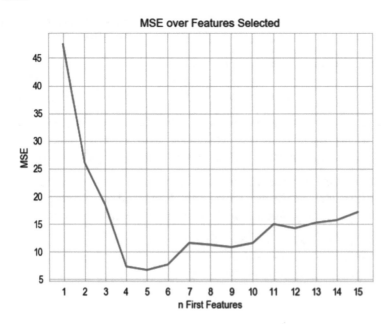

Figure 7.6: Plot of error values for the Friedman-1 regression problem

In the next subsection, we will find out if a genetic algorithm can successfully identify these first five features.

Genetic algorithms solution

To identify the best set of features to be used for our regression test using a genetic algorithm, we've created the Python program, `01_solve_friedman.py`, which is located at `https://github.com/PacktPublishing/Hands-On-Genetic-Algorithms-with-Python-Second-Edition/blob/main/chapter_07/01_solve_friedman.py`.

As a reminder, the chromosome representation that's being used here is a list of integers with values of 0 or 1, denoting whether a feature should be used or dropped. This makes our problem, from the point of view of the genetic algorithm, similar to the *OneMax* problem, or the *knapsack 0-1* problem we solved previously. The difference is in the fitness function returning the regression model's MSE, which is calculated within the `Friedman1Test` class.

The following steps describe the main parts of our solution:

1. First, we need to create an instance of the `Friedman1Test` class with the desired parameters:

```
friedman = friedman.Friedman1Test(NUM_OF_FEATURES, \
    NUM_OF_SAMPLES, RANDOM_SEED)
```

2. Since our goal is to minimize the MSE of the regression model, we define a single objective, minimizing the fitness strategy:

```
creator.create("FitnessMin", base.Fitness, weights=(-1.0,))
```

3. Since the solution is represented by a list of 0 or 1 integer values, we use the following toolbox definitions to create the initial population:

```
toolbox.register("zeroOrOne", random.randint, 0, 1)
toolbox.register("individualCreator",\
    tools.initRepeat, creator.Individual, \
    toolbox.zeroOrOne, len(friedman))
toolbox.register("populationCreator", tools.initRepeat, \
    list, toolbox.individualCreator)
```

4. Then, we instruct the genetic algorithm to use the `getMSE()` method of the `Friedman1Test` instance for fitness evaluation:

```
def friedmanTestScore(individual):
    return friedman.getMSE(individual),  # return a tuple
toolbox.register("evaluate", friedmanTestScore)
```

5. As for the genetic operators, we use *tournament selection* with a tournament size of 2 and *crossover* and *mutation* operators that are specialized for binary list chromosomes:

```
toolbox.register("select", tools.selTournament, tournsize=2)
toolbox.register("mate", tools.cxTwoPoint)
toolbox.register("mutate", tools.mutFlipBit, \
    indpb=1.0/len(friedman))
```

6. In addition, we continue to use the *elitist approach*, where the **hall of fame (HOF)** members – the current best individuals – are always passed untouched to the next generation:

```
population, logbook = elitism.eaSimpleWithElitism(
    population,
    toolbox,
    cxpb=P_CROSSOVER,
    mutpb=P_MUTATION,
    ngen=MAX_GENERATIONS,
    stats=stats,
    halloffame=hof,
    verbose=True)
```

By running the algorithm for 30 generations with a population size of 30, we get the following outcome:

```
-- Best Ever Individual = [1, 1, 1, 1, 1, 0, 0, 0, 0, 0, 0, 0, 0, 0,
0]
-- Best Ever Fitness = 6.702668910463287
```

This indicates that the first five features have been selected to provide the best MSE (about 6.7) for our test. Note that the genetic algorithm makes no assumptions about the set of features that it was looking for, meaning it did not know that we were looking for a subset of the first *n* features. It simply searched for the best possible subset of features.

In the next section, we will advance from using artificially generated data to an actual dataset, and utilize the genetic algorithm to select the best features for a classification problem.

Selecting the features for classifying the Zoo dataset

The UCI Machine Learning Repository (https://archive.ics.uci.edu/) maintains over 600 datasets as a service to the machine learning community. These datasets can be used for experimentation with various models and algorithms. A typical dataset contains a number of features (inputs) and the desired output, in theform of columns, with a description of their meaning.

In this section, we will use the UCI Zoo dataset (`https://archive.ics.uci.edu/dataset/111/zoo`). This dataset describes 101 different animals using the following 18 features:

No.	Feature Name	Data Type
1	animal name	unique for each instance
2	hair	boolean
3	feathers	boolean
4	eggs	boolean
5	milk	boolean
6	airborne	boolean
7	aquatic	boolean
8	predator	boolean
9	toothed	boolean
10	backbone	boolean
11	breathes	boolean
12	venomous	boolean
13	fins	boolean
14	legs	Numeric (set of values {0,2,4,5,6,8})
15	tail	boolean
16	domestic	boolean
17	catsize	boolean
18	type	numeric (integer values in the range [1..7])

Table 7.1: Feature list for the Zoo dataset

Most features are `Boolean` (value of 1 or 0), indicating the presence or absence of a certain attribute, such as `hair`, `fins`, and so on. The first feature, `animal name`, is just to provide us with added information and does not participate in the learning process.

This dataset is used for testing classification tasks, where the input features need to be mapped into two or more categories/labels. In this dataset, the last feature, called `type`, represents the category and is used as the **output** value. For this value, there are seven categories altogether. A `type` value of 5, for instance, represents the animal category that includes frog, newt, and toad.

To sum this up, a classification model trained with this dataset will use features 2–17 (`hair`, `feathers`, `fins`, and so on) to predict the value of feature 18 (animal `type`).

Once again, we want to use a genetic algorithm to select the features that will give us the best predictions. Let's start by creating a Python class that represents a classifier that's been trained with this dataset.

Python problem representation

To encapsulate the feature selection process for the Zoo dataset classification task, we've created a Python class called Zoo. This class is contained in the zoo.py file, which is located at:

https://github.com/PacktPublishing/Hands-On-Genetic-Algorithms-with-Python-Second-Edition/blob/main/chapter_07/zoo.py

The main parts of this class are highlighted as follows:

1. The __init__() method of the class loads the Zoo dataset from the web while skipping the first feature—animal name—as follows:

    ```
    self.data = read_csv(self.DATASET_URL, header=None,
        usecols=range(1, 18))
    ```

2. Then, it separates the data to input features (first remaining 16 columns) and the resulting category (last column):

    ```
    self.X = self.data.iloc[:, 0:16]
    self.y = self.data.iloc[:, 16]
    ```

3. Instead of just separating the data into a training set and a test set, like we did in the previous section, we're using **k-fold cross-validation**. This means that the data is split into *k* equal parts and the model is evaluated *k* times, each time using *(k-1)* parts for training and the remaining part for testing (or *validation*). This is easy to do in Python using the scikit-learn library's model_selection.KFold() method:

    ```
    self.kfold = model_selection.KFold(
        n_splits=self.NUM_FOLDS,
        random_state=self.randomSeed)
    ```

4. Next, we create a classification model based on a **decision tree**. This type of classifier creates a tree structure during the training phase that splits the dataset into smaller subsets, eventually resulting in a prediction:

    ```
    self.classifier = DecisionTreeClassifier(
        random_state=self.randomSeed)
    ```

> **Important Note**
>
> We are passing a random seed so that it can be used internally by the classifier. This way, we can make sure the results that are obtained are repeatable.

5. The `getMeanAccuracy()` method of the class is used to evaluate the performance of the classifier for a set of selected features. Similar to the `getMSE()` method in the `Friedman1Test` class, this method accepts a list of binary values corresponding to the features in the dataset—a value of 1 represents selecting the corresponding feature, while a value of 0 means that the feature is dropped. The method then drops the columns in the dataset that correspond to the unselected features:

```
zeroIndices = [i for i, n in enumerate(zeroOneList) if n == 0]
currentX = self.X.drop(self.X.columns[zeroIndices], axis=1)
```

6. This modified dataset—containing only the selected features—is then used to perform the **k-fold cross-validation** process and determine the classifier's performance over the data partitions. The value of `k` in our class is set to 5, so five evaluations take place each time:

```
cv_results = model_selection.cross_val_score(
    self.classifier, currentX, self.y, cv=self.kfold,
    scoring='accuracy')
return cv_results.mean()
```

The metric that's being used here to evaluate the classifier is accuracy—the portion of the cases that were classified correctly. An accuracy of 0.85, for example, means that 85% of the cases were classified correctly. Since, in our case, we train and evaluate the classifier k times, we use the average (mean) accuracy value that was obtained over these evaluations.

7. The main method of the class creates an instance of the `Zoo` class and evaluates the classifier with all 16 features that are present using the all-one solution representation:

```
allOnes = [1] * len(zoo)
print("-- All features selected: ", allOnes, ", accuracy = ",
    zoo.getMeanAccuracy(allOnes))
```

When running the main method of the class, the printout shows that, after testing our classifier with 5-fold cross-validation using all 16 features, the classification accuracy that's achieved is about 91%:

```
-- All features selected:  [1, 1, 1, 1, 1, 1, 1, 1, 1, 1, 1, 1, 1, 1,
1, 1], accuracy =  0.9099999999999999
```

In the next subsection, we will attempt to improve the accuracy of the classifier by selecting a subset of features from the dataset, instead of using all the features. We will use—you guessed it—a genetic algorithm to select these features for us.

Genetic algorithms solution

To identify the best set of features to be used for our Zoo classification task using a genetic algorithm, we've created the Python program `02_solve_zoo.py`, which is located at `https://github.com/PacktPublishing/Hands-On-Genetic-Algorithms-with-Python-Second-Edition/blob/main/chapter_07/02_solve_zoo.py`. As in the previous section, the chromosome representation that's being used here is a list of integers with the values of 0 or 1, denoting whether a feature should be used or dropped.

The following steps highlight the main parts of the program:

1. First, we need to create an instance of the `Zoo` class and pass our random seed along for the sake of producing repeatable results:

    ```
    zoo = zoo.Zoo(RANDOM_SEED)
    ```

2. Since our goal is to maximize the accuracy of the classifier model, we define a single objective, maximizing the fitness strategy:

    ```
    creator.create("FitnessMax", base.Fitness, weights=(1.0,))
    ```

3. Just like in the previous section, we use the following toolbox definitions to create the initial population of individuals, each constructed as a list of 0 or 1 integer values:

    ```
    toolbox.register("zeroOrOne", random.randint, 0, 1)
    toolbox.register("individualCreator", tools.initRepeat, \
        creator.Individual, toolbox.zeroOrOne, len(zoo))
    toolbox.register("populationCreator", tools.initRepeat, \
        list, toolbox.individualCreator)
    ```

4. Then, we instruct the genetic algorithm to use the `getMeanAccuracy()` method of the `Zoo` instance for fitness evaluation. To do this, we have to make two modifications:

    -   We eliminate the possibility of no features being selected (all-zeros individual) since our classifier will throw an exception in such a case.

    -   We add a small *penalty* for each feature being used to encourage the selection of fewer features. The penalty value is very small (0.001), so it only comes into play as a tie-breaker between two equally performing classifiers, leading the algorithm to prefer the one that uses fewer features:

    ```
    def zooClassificationAccuracy(individual):
        numFeaturesUsed = sum(individual)
        if numFeaturesUsed == 0:
            return 0.0,
        else:
            accuracy = zoo.getMeanAccuracy(individual)
    ```

```
        return accuracy - FEATURE_PENALTY_FACTOR *
            numFeaturesUsed,  # return a tuple
    toolbox.register("evaluate", zooClassificationAccuracy)
```

5. For the genetic operators, we again use **tournament selection** with a tournament size of 2 and **crossover** and **mutation** operators that are specialized for binary list chromosomes:

```
    toolbox.register("select", tools.selTournament, tournsize=2)
    toolbox.register("mate", tools.cxTwoPoint)
    toolbox.register("mutate", tools.mutFlipBit, indpb=1.0/len(zoo))
```

6. And once again, we continue to use the **elitist approach**, where HOF members—the current best individuals—are always passed untouched to the next generation:

```
    population, logbook = elitism.eaSimpleWithElitism(population,
        toolbox,
        cxpb=P_CROSSOVER,
        mutpb=P_MUTATION,
        ngen=MAX_GENERATIONS,
        stats=stats,
        halloffame=hof,
        verbose=True)
```

7. At the end of the run, we print out all the members of the HOF so that we can see the top results that were found by the algorithm. We print both the fitness value, which includes the penalty for the number of features, and the actual accuracy value:

```
    print("- Best solutions are:")
    for i in range(HALL_OF_FAME_SIZE):
        print(
            i, ": ", hof.items[i],
            ", fitness = ", hof.items[i].fitness.values[0],
            ", accuracy = ", zoo.getMeanAccuracy(hof.items[i]),
            ", features = ", sum(hof.items[i])
        )
```

By running the algorithm for 50 generations with a population size of 50 and HOF size of 5, we get the following outcome:

```
- Best solutions are:
0 : [0, 1, 0, 1, 1, 0, 0, 0, 1, 0, 0, 1, 0, 1, 0, 0] , fitness = 0.964
, accuracy = 0.97 , features = 6
1 : [0, 1, 0, 1, 1, 0, 0, 0, 1, 0, 0, 1, 0, 1, 0, 1] , fitness = 0.963
, accuracy = 0.97 , features = 7
2 : [0, 1, 0, 1, 1, 0, 0, 0, 1, 0, 1, 1, 0, 1, 0, 0] , fitness = 0.963
, accuracy = 0.97 , features = 7
```

```
3 : [1, 1, 0, 1, 1, 0, 0, 0, 1, 0, 0, 1, 0, 1, 0, 0] , fitness = 0.963
, accuracy = 0.97 , features = 7
4 : [0, 1, 0, 1, 1, 0, 0, 0, 1, 0, 0, 1, 0, 1, 1, 0] , fitness = 0.963
, accuracy = 0.97 , features = 7
```

These results indicate that all five top solutions achieved an accuracy value of 97%, using either six or seven features out of the available 16. Thanks to the penalty factor on a number of features, the top solution is the set of six features, which are as follows:

- `feathers`
- `milk`
- `airborne`
- `backbone`
- `fins`
- `tail`

In conclusion, by selecting these particular features out of the 16 given in the dataset, not only did we reduce the dimensionality of the problem, but we were also able to improve our model's accuracy from 91% to 97%. If this does not seem like a large enhancement at first glance, think of it as reducing the error rate from 9% to 3% – a very significant improvement in terms of classification performance.

Summary

In this chapter, you were introduced to machine learning and the two main types of supervised machine learning tasks – *regression* and *classification*. Then, you were presented with the potential benefits of *feature selection* on the performance of the models carrying out these tasks. At the heart of this chapter were two demonstrations of how genetic algorithms can be utilized to enhance the performance of such models via feature selection. In the first case, we pinpointed the genuine features that were generated by the *Friedman-1 Test* regression problem, while, in the other case, we selected the most beneficial features of the *Zoo classification dataset*.

In the next chapter, we will look at another possible way of enhancing the performance of supervised machine learning models, namely **hyperparameter tuning**.

Further reading

For more information about the topics that were covered in this chapter, please refer to the following resources:

- *Applied Supervised Learning with Python*, Benjamin Johnston and Ishita Mathur, April 26, 2019

- *Feature Engineering Made Easy*, Sinan Ozdemir and Divya Susarla, January 22, 2018

- *Feature selection for classification*, M.Dash and H.Liu, 1997: `https://doi.org/10.1016/S1088-467X(97)00008-5`

- *UCI Machine Learning Repository*: `https://archive.ics.uci.edu/`

8

Hyperparameter Tuning of Machine Learning Models

This chapter describes how genetic algorithms can be used to improve the performance of **supervised machine learning** models by tuning the hyperparameters of the models. The chapter will start with a brief introduction to **hyperparameter tuning** in machine learning before describing the concept of a **grid search**. After introducing the Wine dataset and the adaptive boosting classifier, both of which will be used throughout this chapter, we will demonstrate hyperparameter tuning using both a conventional grid search and a genetic-algorithm-driven grid search. Finally, we will attempt to enhance the results we get by using a direct genetic algorithm approach for hyperparameter tuning.

By the end of this chapter, you will be able to do the following:

- Demonstrate familiarity with the concept of hyperparameter tuning in machine learning
- Demonstrate familiarity with the Wine dataset and the adaptive boosting classifier
- Enhance the performance of a classifier using a hyperparameter grid search
- Enhance the performance of a classifier using a genetic-algorithm-driven hyperparameter grid search
- Enhance the performance of a classifier using a direct genetic algorithm approach for hyperparameter tuning

We will start this chapter with a quick overview of hyperparameters in machine learning. If you are a seasoned data scientist, feel free to skip the introductory section.

Technical requirements

In this chapter, we will be using Python 3 with the following supporting libraries:

- `deap`
- `numpy`
- `pandas`
- `matplotlib`
- `seaborn`
- `scikit-learn`

> **Important note**
>
> If you use the `requirements.txt` file we provide (see *Chapter 3*), these libraries are already included in your environment.

In addition, we will be using the UCI Wine dataset: `https://archive.ics.uci.edu/ml/datasets/Wine`

The programs that will be used in this chapter can be found in this book's GitHub repository:

`https://github.com/PacktPublishing/Hands-On-Genetic-Algorithms-with-Python-Second-Edition/tree/main/chapter_08`

Check out the following video to see the code in action: `https://packt.link/OEBOd`

Hyperparameters in machine learning

In *Chapter 7, Enhancing Machine Learning Models Using Feature Selection*, we described *supervised learning* as the programmatic process of adjusting (or tuning) the internal parameters of a model to produce the desired outputs in response to given inputs. To make this happen, each type of supervised learning model is accompanied by a learning algorithm that iteratively adjusts its internal parameters during the *learning* (or training) phase.

However, most models have another set of parameters that are set *before* the learning takes place. These are called **hyperparameters** and affect the way the learning is done. The following figure illustrates the two types of parameters:

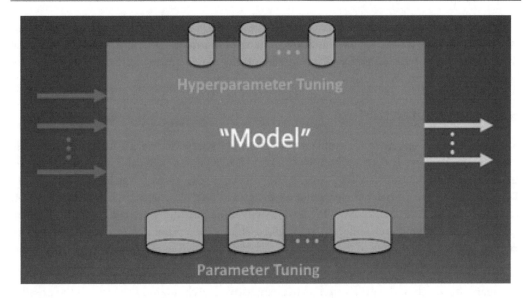

Figure 8.1: Hyperparameter tuning of a machine learning model

Usually, the hyperparameters have default values that will take effect if we don't specifically set them. For example, if we look at the `scikit-learn` library implementation of the **decision tree classifier** (`https://scikit-learn.org/stable/modules/generated/sklearn.tree.DecisionTreeClassifier.html`), we will see several hyperparameters and their default values.

A few of these hyperparameters are described in the following table:

Name	Type	Description	Default value
max_depth	int	The maximum depth of the tree	None
splitter	enumerated	The strategy that's used to choose the split at each best node: `{'best', 'random'}`	`'best'`
min_samples_split	int or float	The minimum number of samples required to split an internal node	2

Table 8.1: Hyperparameters and their details

Each of these parameters affects the way the decision tree is constructed during the learning process, and their combined effect on the results of the learning process—and, consequently, on the performance of the model—can be significant.

Since the choice of hyperparameters has a considerable impact on the performance of machine learning models, data scientists often spend significant amounts of time looking for the best hyperparameter combinations, a process called **hyperparameter tuning**. Some of the methods that are used for hyperparameter tuning will be described in the next subsection.

Hyperparameter tuning

A common way of searching for good combinations of hyperparameters is using a **grid search**. Using this method, we choose a subset of values for each hyperparameter that we want to tune. As an example, given the Decision Tree classifier, we can choose the subset of values, such as {2, 5, 10}, for the `max_depth` parameter, while, for the `splitter` parameter, we choose both possible values—{"best", "random"}. Then, we try out all six possible combinations of these values. For each combination, the classifier is trained and evaluated for a certain performance criterion; for example, accuracy. At the end of the process, we pick the combination of hyperparameter values that yielded the best performance.

The main drawback of the grid search is the exhaustive search it conducts over all the possible combinations, which can prove very lengthy. One common way to produce good combinations in a shorter amount of time is **random search**, where random combinations of hyperparameters are chosen and tested.

A better option—of particular interest to us—when it comes to performing the grid search is harnessing a genetic algorithm to look for the best combination(s) of hyperparameters within the predefined grid. This method offers the potential for finding the best grid combinations in a shorter amount of time than the original, exhaustive grid search.

While grid search and random search are supported by the `scikit-learn` library, a genetic algorithm-driven grid search option is offered by `sklearn-deap`. This small library builds upon the DEAP-based genetic algorithm's capabilities, as well as the existing features of `scikit-learn`. At the time of writing this book, this library is not in sync with the latest version of `scikit-learn`; therefore, we included a slightly modified version of it under the `sklearn_deap` folder as part of the files of *Chapter 8*; we will make use of that version.

In the following sections, we will compare both approaches to the grid search—exhaustive and genetic-algorithm-driven. But first, we'll take a quick look at the dataset we are going to use for our experiment—the **UCI Wine dataset**.

The Wine dataset

A commonly used dataset from the *UCI Machine Learning Repository* (https://archive.ics.uci.edu/), the Wine dataset (https://archive.ics.uci.edu/ml/datasets/Wine) contains the results of a chemical analysis that was conducted for 178 different wines that were grown in the same region in Italy. These wines are categorized into one of three types.

The chemical analysis consists of 13 different measurements, representing the quantities of the following constituents that are found in each wine:

- Alcohol
- Malic acid
- Ash
- Alkalinity of ash
- Magnesium
- Total phenols
- Flavanoids
- Non-flavanoid phenols
- Proanthocyanins
- Color intensity
- Hue
- OD280/OD315 of diluted wines
- Proline

Columns 2-14 of the dataset contain the values for the preceding measurements, while the classification outcome—the wine type itself (1, 2, or 3)—is found in the first column.

Next, let's look at the classifier we chose to classify this dataset.

The adaptive boosting classifier

The **adaptive boosting algorithm**, or **AdaBoost**, for short, is a powerful machine learning model that combines the outputs of multiple instances of a simple learning algorithm (**weak learner**) using a weighted sum. AdaBoost adds instances of the weak learner during the learning process, each of which is adjusted to improve previously misclassified inputs.

The `scikit-learn` library's implementation of this model, the Adaboost classifier (`https://scikit-learn.org/stable/modules/generated/sklearn.ensemble.AdaBoostClassifier.html`), uses several hyperparameters, some of which are as follows:

Name	Type	Description	Default value
`n_estimators`	int	The maximum number of estimators	`50`
`learning_rate`	float	Weight applied to each classifier at each boosting iteration; a higher learning rate increases the contribution of each classifier	`1.0`
`algorithm`	enumerated	The boosting algorithm to be used: {`'SAMME'` , `'SAMME.R'`}	`'SAMME.R'`

Table 8.1: Hyperparameters and their details

Interestingly, each of these three hyperparameters is of a different type—an int, a float, and an enumerated (or categorical) type. Later, we will find out how each tuning method handles these different types. We will start with two forms of the grid search, both of which will be described in the next section.

Tuning the hyperparameters using conventional versus genetic grid search

To encapsulate the hyperparameter tuning of the AdaBoost classifier for the Wine dataset using a grid search—both the conventional version and the genetic-algorithm-driven version—we created a Python class called `HyperparameterTuningGrid`. This class can be found in the `01_hyperparameter_tuning_grid.py` file, which is located at

`https://github.com/PacktPublishing/Hands-On-Genetic-Algorithms-with-Python-Second-Edition/blob/main/chapter_08/01_hyperparameter_tuning_grid.py`.

The main parts of this class are highlighted as follows:

1. The `__init__()` method of the class initializes the wine dataset, the AdaBoost classifier, the k-fold cross-validation metric, and the grid parameters:

    ```
    self.initWineDataset()
    self.initClassifier()
    self.initKfold()
    self.initGridParams()
    ```

2. The `initGridParams()` method initializes the grid search by setting the tested values of the three hyperparameters mentioned in the previous section:

    -   The `n_estimators` parameter is tested across 10 values, linearly spaced between 10 and 100.

    -   The `learning_rate` parameter is tested across 100 values, logarithmically spaced between 0.1 (10^{-2}) and 1 (10^0).

    -   Both possible values of the `algorithm` parameter, `'SAMME'` and `'SAMME.R'`, are tested.

    This setup covers a total of 200 (10×10×2) different combinations of the grid parameters:

    ```
    self.gridParams = {
        'n_estimators': [10, 20, 30, 40, 50, 60, 70, 80, 90, 100],
        'learning_rate': np.logspace(-2, 0, num=10, base=10),
        'algorithm': ['SAMME', 'SAMME.R'],
    }
    ```

3. The `getDefaultAccuracy()` method evaluates the accuracy of the classifier with its default hyperparameter values using the mean value of the `'accuracy'` metric:

    ```
    cv_results = model_selection.cross_val_score(
        self.classifier,
        self.X,
        self.y,
        cv=self.kfold,
        scoring='accuracy')
    return cv_results.mean()
    ```

4. The `gridTest()` method performs a conventional grid search over the set of tested hyperparameter values we defined earlier. The best combination of parameters is determined based on the k-fold cross-validation mean `'accuracy'` metric:

    ```
    gridSearch = GridSearchCV(
        estimator=self.classifier,
        param_grid=self.gridParams,
        cv=self.kfold,
        scoring='accuracy')
    gridSearch.fit(self.X, self.y)
    ```

5. The `geneticGridTest()` method performs a genetic-algorithm-driven grid search. It utilizes the `sklearn-deap` library's `EvolutionaryAlgorithmSearchCV()` method, which was designed to be called in a very similar manner to that of the conventional grid search. All we need to do is add a few genetic algorithm parameters—population size, mutation probability, tournament size, and the number of generations:

    ```
    gridSearch = EvolutionaryAlgorithmSearchCV(
        estimator=self.classifier,
    ```

```
            params=self.gridParams,
            cv=self.kfold,
            scoring='accuracy',
            verbose=True,
            population_size=20,
            gene_mutation_prob=0.50,
            tournament_size=2,
            generations_number=5)
    gridSearch.fit(self.X, self.y)
```

6. Finally, the `main()` method of the class starts by evaluating the performance of the classifier with its default hyperparameter values. Then, it runs the conventional, exhaustive grid search, followed by the genetic-algorithm-driven grid search, while timing each search.

The results of running the main method of this class are described in the next subsection.

Testing the classifier's default performance

The results of the run indicate that, with the default parameter values of `n_estimators = 50`, `learning_rate = 1.0`, and `algorithm = 'SAMME.R'`, the classification accuracy of the classifier is about 66.4%:

```
Default Classifier Hyperparameter values:
{'algorithm': 'SAMME.R', 'base_estimator': 'deprecated', 'estimator':
None, 'learning_rate': 1.0, 'n_estimators': 50, 'random_state': 42}
score with default values =  0.6636507936507937
```

This is not a particularly good accuracy. Hopefully, grid search can improve this by finding a better combination of hyperparameter values.

Running the conventional grid search

The conventional, exhaustive grid search, covering all 200 possible combinations, is performed next. The search results indicated that the best combination within this grid was `n_estimators = 50`, `learning_rate ≈ 0.5995`, and `algorithm = 'SAMME.R'`.

The classification accuracy that we achieved with these values is about 92.7%, which is a vast improvement over the original 66.4%. The search runtime was about 131 seconds using a relatively old computer:

```
performing grid search...
best parameters:  {'algorithm': 'SAMME.R', 'learning_rate':
0.5994842503189409, 'n_estimators': 50}
best score:  0.9266666666666667
Time Elapsed =  131.01380705833435
```

Next comes the genetic-powered grid search. Will it match these results? Let's find out.

Running the genetic-algorithm-driven grid search

The last portion of the run describes the genetic-algorithm-driven grid search, which is carried out with the same grid parameters. The verbose output of the search starts with a somewhat cryptic printout:

```
performing Genetic grid search...
Types [1, 2, 1] and maxint [9, 9, 1] detected
```

This printout describes the grid we are searching on—a list of 10 integers (n_estimators values), an ndarray of 10 elements (learning_rate values), and a list of two strings (algorithm values)—as follows:

- Types [1, 2, 1] refers to the grid types of [list, ndarray, list]
- maxint [9, 9, 1] corresponds to the list/array sizes of [10, 10, 2]

The next printed line refers to the total amount of possible grid combinations (10×10×2):

```
--- Evolve in 200 possible combinations ---
```

The rest of the printout looks very familiar since it utilizes the same DEAP-based genetic algorithm tools that we have been using all along, detailing the process of evolving the generations and printing a statistics line for each generation:

gen	nevals	avg	min	max	std
0	20	0.708146	0.117978	0.910112	0.265811
1	13	0.870787	0.662921	0.910112	0.0701235
2	10	0.857865	0.662921	0.91573	0.0735955
3	12	0.87809	0.679775	0.904494	0.0473746
4	12	0.878933	0.662921	0.910112	0.0524153
5	7	0.864045	0.162921	0.926966	0.161174

At the end of the process, the best combination is printed, along with the score value and the time that elapsed:

```
Best individual is: {'n_estimators': 50, 'learning_rate':
0.5994842503189409, 'algorithm': 'SAMME.R'}
with fitness: 0.9269662921348315
Time Elapsed = 21.147947072982788
```

These results indicate that the genetic-algorithm-driven grid search was able to find the same best result that was found using the exhaustive search but in a shorter amount of time.

Please note that this is a simple example that runs very quickly. In real-life situations, we often encounter large datasets, as well as complex models and extensive hyperparameter grids. In these circumstances, running an exhaustive grid search can be prohibitively lengthy, while the genetic-algorithm-driven grid search has the potential to yield good results within a reasonable amount of time.

But still, all grid searches, genetic-driven or not, are limited to the subset of hyperparameter values that are defined by the grid. What if we would like to search outside the grid without being limited to a subset of predefined values? A possible solution is described in the next section.

Tuning the hyperparameters using a direct genetic approach

Besides offering an efficient grid search option, genetic algorithms can be utilized to directly search the entire parameter space, just as we used them to search the input space for many types of problems throughout this book. Each hyperparameter can be represented as a variable participating in the search, and the chromosome can be a combination of all these variables.

Since the hyperparameters can be of varying types—for example, the types float, int, and enumerated, which we have in our AdaBoost classifier—we may want to code each of them differently and then define the genetic operations as a combination of separate operators that are adapted to each of the types. However, we can also use a lazy approach and code all of them as float parameters to simplify the implementation of the algorithm, as we will see next.

Hyperparameter representation

In *Chapter 6, Optimizing Continuous Functions*, we used genetic algorithms to optimize the functions of real-valued parameters. These parameters were represented as a list of float numbers: *[1.23, 7.2134, -25.309]*.

Consequently, the genetic operators we used were specialized for handling lists of floating-point numbers.

To adapt this approach so that it can tune the hyperparameters, we are going to represent each hyperparameter as a floating-point number, regardless of its actual type. To make this work, we need to find a way to transform each parameter into a floating-point number and back from a floating-point number to its original representation. We will implement the following transformations:

- `n_estimators`, originally an integer, will be represented by a float value in a certain range; for example, `[1, 100]`. To transform the float value back into an integer, we will use the Python `round()` function, which will round it to the nearest integer.

- `learning_rate` is already a float, so no conversion is needed. It will be bound to the range of `[0.01, 1.0]`.

- `algorithm` can have one of two values, `'SAMME'` or `'SAMME.R'`, and will be represented by a float number in the range of `[0, 1]`. To transform the float value, we will round it to the nearest integer—0 or 1. Then, we will replace a value of 0 with `'SAMME'` and a value of 1 with `'SAMME.R'`.

These conversions will be carried out by two Python files, both of which will be described in the following subsections.

Evaluating the classifier accuracy

We start with a Python class encapsulating the classifier's *accuracy* evaluation, called `HyperparameterTuningGenetic`. This class can be found in the `hyperparameter_tuning_genetic_test.py` file, which is located at

`https://github.com/PacktPublishing/Hands-On-Genetic-Algorithms-with-Python-Second-Edition/blob/main/chapter_08/hyperparameter_tuning_genetic_test.py`.

The main functionality of this class is highlighted as follows:

1. The `convertParam()` method of the class takes a list called `params`, containing the float values representing the hyperparameters, and transforms them into their actual values (as discussed in the previous subsection):

    ```
    n_estimators = round(params[0])
    learning_rate = params[1]
    algorithm = ['SAMME', 'SAMME.R'][round(params[2])]
    ```

2. The `getAccuracy()` method takes a list of float numbers representing the hyperparameter values, uses the `convertParam()` method to transform them into actual values, and initializes the AdaBoost classifier with these values:

    ```
    n_estimators, learning_rate, algorithm = \
        self.convertParams(params)
    self.classifier = AdaBoostClassifier(
        n_estimators=n_estimators,
        learning_rate=learning_rate,
        algorithm=algorithm)
    ```

3. Then, it finds the accuracy of the classifier using the k-fold cross-validation code that we created for the wine dataset:

    ```
    cv_results = model_selection.cross_val_score(
        self.classifier,
        self.X,
        self.y,
        cv=self.kfold,
        scoring='accuracy')
    return cv_results.mean()
    ```

This class is utilized by the program that implements the hyperparameter-tuning genetic algorithm, as will be described in the next section.

Tuning the hyperparameters using genetic algorithms

The genetic-algorithm-based search for the best hyperparameter values is implemented by the Python program, `02_hyperparameter_tuning_genetic.py`, which is located at

`https://github.com/PacktPublishing/Hands-On-Genetic-Algorithms-with-Python-Second-Edition/blob/main/chapter_08/02_hyperparameter_tuning_genetic.py`.

The following steps describe the main parts of this program:

1. We start by setting the lower and upper boundary for each of the float values representing a hyperparameter, as described in the previous subsection—`[1, 100]` for `n_estimators`, `[0.01, 1]` for `learning_rate`, and `[0, 1]` for `algorithm`:

    ```
    # [n_estimators, learning_rate, algorithm]:
    BOUNDS_LOW  = [  1, 0.01, 0]
    BOUNDS_HIGH = [100, 1.00, 1]
    ```

2. Then, we create an instance of the `HyperparameterTuningGenetic` class that will allow us to test the various combinations of the hyperparameters:

    ```
    test = HyperparameterTuningGenetic(RANDOM_SEED)
    ```

3. Since our goal is to maximize the accuracy of the classifier, we define a single objective, maximizing fitness strategy:

    ```
    creator.create("FitnessMax", base.Fitness, weights=(1.0,))
    ```

4. Now comes a particularly interesting part—since the solution is represented by a list of float values, each of a different range, we use the following loop to iterate over all pairs of lower-bound and upper-bound values. For each hyperparameter, we create a separate toolbox operator, which will be used to generate random float values in the appropriate range:

    ```
    for i in range(NUM_OF_PARAMS):
        # "hyperparameter_0", "hyperparameter_1", ...
        toolbox.register("hyperparameter_" + str(i),
                         random.uniform,
                         BOUNDS_LOW[i],
                         BOUNDS_HIGH[i])
    ```

5. Then, we create the hyperparameter tuple, which contains the specific float number generators we just created for each hyperparameter:

    ```
    hyperparameters = ()
    for i in range(NUM_OF_PARAMS):
    ```

```
hyperparameters = hyperparameters + \
    (toolbox.__getattribute__("hyperparameter_" + str(i)),)
```

6. Now, we can use this hyperparameter tuple, in conjunction with DEAP's built-in `initCycle()` operator, to create a new `individualCreator` operator that fills up an individual instance with a combination of randomly generated hyperparameter values:

```
toolbox.register("individualCreator",
                 tools.initCycle,
                 creator.Individual,
                 hyperparameters,
                 n=1)
```

7. Then, we instruct the genetic algorithm to use the `getAccuracy()` method of the `HyperparameterTuningGenetic` instance for fitness evaluation. As a reminder, the `getAccuracy()` method, which we described in the previous subsection, converts the given individual—a list of three floats—back into the classifier hyperparameter values they represent, trains the classifier with these values, and evaluates its accuracy using k-fold cross-validation:

```
def classificationAccuracy(individual):
    return test.getAccuracy(individual),
toolbox.register("evaluate", classificationAccuracy)
```

8. Now, we need to define the genetic operators. While for the `selection` operator, we use the usual tournament selection with a tournament size of 2, we choose `crossover` and `mutation` operators that are specialized for bounded float-list chromosomes and provide them with the boundaries we defined for each hyperparameter:

```
toolbox.register("select", tools.selTournament, tournsize=2)
toolbox.register("mate",
                 tools.cxSimulatedBinaryBounded,
                 low=BOUNDS_LOW,
                 up=BOUNDS_HIGH,
                 eta=CROWDING_FACTOR)
toolbox.register("mutate",
                 tools.mutPolynomialBounded,
                 low=BOUNDS_LOW,
                 up=BOUNDS_HIGH,
                 eta=CROWDING_FACTOR,
                 indpb=1.0 / NUM_OF_PARAMS)
```

9. In addition, we continue to use the elitist approach, where the HOF members—the current best individuals—are always passed untouched to the next generation:

```
population, logbook = elitism.eaSimpleWithElitism(
    population,
    toolbox,
    cxpb=P_CROSSOVER,
    mutpb=P_MUTATION,
    ngen=MAX_GENERATIONS,
    stats=stats,
    halloffame=hof,
    verbose=True)
```

By running the algorithm for five generations with a population size of 30, we get the following outcome:

```
gen nevals max avg
0       30      0.927143        0.831439
1       22      0.93254         0.902741
2       23      0.93254         0.907847
3       25      0.943651        0.916566
4       24      0.943651        0.921106
5       24      0.943651        0.921751
- Best solution is:
params =    'n_estimators'= 30, 'learning_rate'=0.613,
'algorithm'=SAMME.R
Accuracy = 0.94365
```

These results indicate that the best combination that was found was `n_estimators` = 30, `learning_rate` = `0.613`, and `algorithm` = `'SAMME.R'`.

The classification accuracy that we achieved with these values is about 94.4%—a worthy improvement over the accuracy we achieved with the grid search. Interestingly, the best value that was found for `learning_rate` is just outside the grid values we searched on.

Dedicated libraries

In recent years, several genetic-algorithm-based libraries have been developed that are dedicated to optimizing machine learning model development. One of them is `sklearn-genetic-opt` (`https://sklearn-genetic-opt.readthedocs.io/en/stable/index.html`); it supports both hyperparameters tuning and feature selection. Another more elaborate library is TPOT (`https://epistasislab.github.io/tpot/`); this library provides optimization for the end-to-end machine learning development process, also called the **pipeline**. You are encouraged to try out these libraries in your own projects.

Summary

In this chapter, you were introduced to the concept of hyperparameter tuning in machine learning. After getting acquainted with the Wine dataset and the AdaBoost classifier, both of which we used for testing throughout this chapter, you were presented with the hyperparameter tuning methods of an exhaustive grid search and its genetic-algorithm-driven counterpart. These two methods were then compared using our test scenario. Finally, we tried out a direct genetic algorithm approach, where all the hyperparameters were represented as float values. This approach allowed us to improve the results of the grid search.

In the next chapter, we will look into the fascinating machine learning models of **neural networks** and **deep learning** and apply genetic algorithms to improve their performance.

Further reading

For more information on the topics that were covered in this chapter, please refer to the following resources:

- Cross-validation and Parameter Tuning, from the book *Mastering Predictive Analytics with scikit-learn and TensorFlow*, Alan Fontaine, September 2018:
- `https://subscription.packtpub.com/book/big_data_and_business_intelligence/9781789617740/2/ch02lvl1sec16/introduction-to-hyperparameter-tuning`
- *sklearn-deap* at GitHub: `https://github.com/rsteca/sklearn-deap`
- *Scikit-learn* AdaBoost Classifier: `https://scikit-learn.org/stable/modules/generated/sklearn.ensemble.AdaBoostClassifier.html`
- *UCI Machine Learning Repository*: `https://archive.ics.uci.edu/`

9

Architecture Optimization of Deep Learning Networks

This chapter describes how genetic algorithms can be used to improve the performance of **artificial neural network** (**ANN**)-based models by optimizing the **network architecture** of these models. We will start with a brief introduction to **neural networks** (**NNs**) and **deep learning** (**DL**). After introducing the *Iris dataset* and **Multilayer Perceptron** (**MLP**) classifiers, we will demonstrate **network architecture optimization** using a genetic algorithm-based solution. Then, we will extend this approach to combine network architecture optimization with model **hyperparameter tuning**, which will be jointly carried out by a genetic algorithm-based solution.

In this chapter, we will cover the following topics:

- Understanding the basic concepts of ANNs and DL

- Enhancing the performance of a DL classifier using network architecture optimization

- Further enhancing the performance of the DL classifier by combining network architecture optimization with hyperparameter tuning

We will start this chapter with an overview of ANNs. If you are a seasoned data scientist, feel free to skip the introductory sections.

Technical requirements

In this chapter, we will be using Python 3 with the following supporting libraries:

- `deap`

- `numpy`

- `scikit-learn`

> **Important note**
>
> If you use the `requirements.txt` file we provide (see *Chapter 3*), these libraries are already included in your environment.

In addition, we will be using the UCI Iris flower dataset (`https://archive.ics.uci.edu/ml/datasets/Iris`).

The programs that will be used in this chapter can be found in this book's GitHub repository at the following link:

`https://github.com/PacktPublishing/Hands-On-Genetic-Algorithms-with-Python-Second-Edition/tree/main/chapter_09`

Check out the following video to see the code in action: `https://packt.link/OEBOd`

ANNs and DL

Inspired by the structure of the human brain, NNs are among the most commonly used models in **machine learning** (**ML**). The basic building blocks of these networks are nodes, or **neurons**, which are based on the biological neuron cell, as depicted in the following diagram:

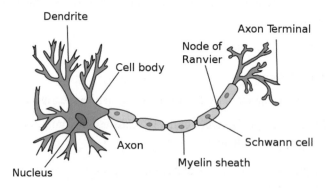

Figure 9.1: Biological neuron model

Source: `https://simple.wikipedia.org/wiki/Neuron#/media/File:Neuron.svg` by Dhp1080

The neuron cell's **dendrites**, which surround the **cell body** on the left-hand side of the preceding diagram, are used as inputs from multiple similar cells, while the long **axon**, coming out of the **cell body,** serves as output and can be connected to multiple other cells via its **terminals**.

This structure is mimicked by an artificial model called a **perceptron**, illustrated as follows:

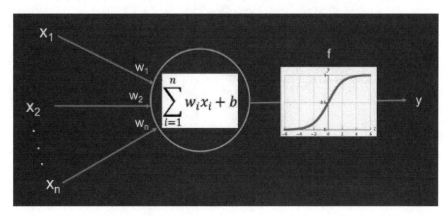

Figure 9.2: Artificial neuron model – the perceptron

The perceptron calculates the output by multiplying each of the input values by a certain **weight**; the results are accumulated, and a **bias** value is added to the sum. A non-linear **activation function** then maps the result to the output. This functionality emulates the operation of the biological neuron, which fires (sends a series of pulses from its output) when the weighted sum of the inputs is above a certain threshold.

The perceptron model can be used for simple classification and regression tasks if we adjust its weight and bias values so that they map certain inputs to the desired output levels. However, a much more capable model can be constructed when connecting multiple perceptron units in a structure called an MLP, which will be described in the next subsection.

MLP

An MLP extends the idea of the perceptron by using numerous nodes, each one implementing a perceptron. The nodes in an MLP are arranged in **layers**, and each layer is connected to the next. The basic structure of an MLP is illustrated in the following diagram:

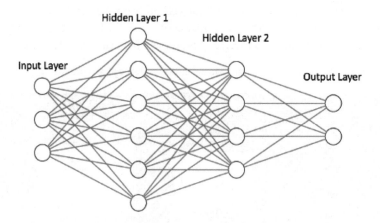

Figure 9.3: The basic structure of an MLP

An MLP consists of three main parts:

- **Input layer**: Receives the input values and connects each of them to every neuron in the next layer.

- **Output layer**: Delivers the results calculated by the MLP. When the MLP is used as a **classifier**, each of the outputs represents one of the classes. When the MLP is used for **regression**, there will be a single output node, producing a continuous value.

- **Hidden layer(s)**: Provide the true power and complexity of this model. While the preceding diagram shows only two hidden layers, there can be numerous hidden layers, each an arbitrary size, that are placed between the input and output layers. As the number of hidden layers grows, the network becomes deeper and is capable of performing an increasingly more complex and non-linear mapping between the inputs and the outputs.

Training this model involves adjusting the weight and bias values for each of the nodes. This is typically done using a family of algorithms called **backpropagation**. The basic principle of backpropagation is to minimize the error between the actual outputs and the desired ones by propagating the output error through the layers of the MLP model, from the output layer inward. The process begins by defining a cost (or "loss") function, typically a measure of the difference between the predicted outputs and the actual target values. The weights and biases of the various nodes are adjusted so that those that contributed the most to the error see the greatest adjustments. By iteratively reducing the cost function, the algorithm refines the model parameters to improve performance.

For many years, the computational limitations of backpropagation algorithms restricted MLPs to no more than two or three hidden layers, until new developments changed matters dramatically. These will be explained in the next section.

DL and convolutional NNs

In recent years, backpropagation algorithms have made a leap forward, enabling the use of a large number of hidden layers in a single network. In these **deep NNs** (**DNNs**), each layer can interpret a combination of several simpler abstract concepts that were learned by the nodes of the previous layer and produce higher-level concepts. For example, when implementing a face recognition task, the first layer will process the pixels of an image and learn to detect edges in different orientations. The next layer may assemble these into lines, corners, and so on, up to a layer that detects facial features such as nose and lips, and finally, one that combines these into the complete concept of a face.

Further advancements have brought about the idea of **convolutional NNs** (**CNNs**). These structures can reduce the count of nodes in DNNs that process two-dimensional information (such as images) by treating nearby inputs differently compared to inputs that are far apart. As a result, these models have proved especially successful when it comes to image and video processing tasks. Besides fully connected layers, similar to the hidden layers in the MLP, these networks utilize pooling (down-sampling) layers, which aggregate outputs of neurons from preceding layers, and convolutional layers, which are used for detecting specific features, such as edges in various orientations, by effectively sliding a filter over the input image.

Training such **DL models** can be computationally intensive and is often done with the aid of **graphics processing units** (**GPUs**), which are more efficient than ordinary CPUs in implementing the backpropagation algorithm. Specialized DL libraries, such as Torch and TensorFlow, are capable of utilizing GPU-based computing platforms. In this chapter, however, for the sake of simplicity, we will be using the MLP implementation offered by the `scikit-learn` library and a simple dataset. The principles that will be used, however, still apply to more complex networks and datasets.

In the next section, we will find out how the architecture of an MLP can be optimized using a genetic algorithm.

Optimizing the architecture of a DL classifier

When creating a NN model for a given ML task, one crucial design decision that needs to be made is the configuration of the **network architecture**. In the case of an MLP, the number of nodes in the input and output layers is determined by the characteristics of the problem at hand. Therefore, the choices to be made are about the hidden layers—how many layers, and how many nodes are in each layer. Some rules of thumb can be employed for making these decisions, but in many cases, identifying the best choices can turn into a cumbersome trial-and-error process.

One way to handle network architecture parameters is to consider them as hyperparameters of the model since they need to be determined before training is done and, consequently, affect the training's results. In this section, we are going to apply this approach and use a genetic algorithm to search for the best combination of hidden layers, in a similar manner to the way we went about choosing the best hyperparameter values in the previous chapter. Let's start with the task we want to tackle – the **Iris flower classification**.

The Iris flower dataset

Perhaps the most well-studied dataset, the *Iris flower dataset* (`https://archive.ics.uci.edu/ml/datasets/Iris`) contains measurements of the `sepal` and `petal` parts of three Iris species (Iris setosa, Iris virginica, and Iris versicolor), as taken by biologists in 1936.

The dataset contains 50 samples from each of the three species, and consists of the following four features:

- `sepal_length (cm)`
- `sepal_width (cm)`
- `petal_length (cm)`
- `petal_width (cm)`

This dataset is directly available via the `scikit-learn` library and can be initialized as follows:

```
from sklearn import datasets
data = datasets.load_iris()
X = data['data']
y = data['target']
```

In our experiments, we will be using an MLP classifier in conjunction with this dataset and harness the power of genetic algorithms to find the network architecture—the number of hidden layers and the number of nodes in each layer—that will yield the best classification accuracy.

Since we are using the genetic algorithms approach, the first thing we need to do is find a way to represent this architecture using a chromosome, as described in the next subsection.

Representing the hidden layer configuration

Since the architecture of an MLP is determined by the hidden layer configuration, let's explore how this configuration can be represented in our solution. The hidden layer configuration of the `sklearn` MLP (`https://scikit-learn.org/stable/modules/neural_networks_supervised.html`) model is conveyed via the `hidden_layer_sizes` tuple, which is sent as a parameter to the model's constructor. By default, the value of this tuple is `(100,)`, which means a single hidden layer of 100 nodes. If we wanted, for example, to configure the MLP with three hidden layers of 20 nodes each, this parameter's value would be `(20, 20, 20)`. Before we implement our genetic algorithm-based optimizer for the hidden layer configuration, we need to define a chromosome that can be translated into this pattern.

To accomplish this, we need to come up with a chromosome that can both express the number of layers and the number of nodes in each layer. A variable-length chromosome that can be directly translated into the variable-length tuple that's used as the model's `hidden_layer_sizes` parameter is one option; however, this approach would require custom, possibly cumbersome, genetic operators. To be able to use our standard genetic operators, we will use a fixed-length representation. When using this approach, the maximum number of layers is decided in advance, and all the layers are always represented, but not necessarily expressed in the solution. For example, if we decide to limit the network to four hidden layers, the chromosome will look as follows:

$$[n_1, n_2, n_3, n_4]$$

Here, n_i denotes the number of nodes in the layer i. However, to control the actual number of hidden layers in the network, some of these values may be zero, or negative. Such a value means that no more layers will be added to the network. The following examples illustrate this method:

- The chromosome [10, 20, -5, 15] is translated into the tuple (10, 20) since -5 terminates the layer count
- The chromosome [10, 0, -5, 15] is translated into the tuple (10,) since 0 terminates the layer count
- The chromosome [10, 20, 5, -15] is translated into the tuple (10, 20, 5) since -15 terminates the layer count
- The chromosome [10, 20, 5, 15] is translated into the tuple (10, 20, 5, 15)

To guarantee that there is at least one hidden layer, we can make sure that the first parameter is always greater than zero. The other parameters can have varying distributions around zero so that we can control their chances of being the terminating parameters.

In addition, even though this chromosome is made up of integers, we chose to utilize float numbers instead, just like we did in the previous chapter for various types of variables. Using a list of float numbers is convenient as it allows us to use existing genetic operators while being able to easily extend the chromosome so that it includes other parameters of different types, which we will do later on. The float numbers can be translated back into integers using the `round()` function. A couple of examples of this generalized approach are as follows:

- The chromosome [9.35, 10.71, -2.51, 17.99] is translated into the tuple (9, 11)
- The chromosome [9.35, 10.71, 2.51, -17.99] is translated into the tuple (9, 11, 3)

To evaluate a given architecture-representing chromosome, we will need to translate it back into the tuple of layers, create an MLP classifier implementing these layers, train it, and evaluate it. We will learn how to do this in the next subsection.

Evaluating the classifier's accuracy

Let's start with a Python class that encapsulates the MLP classifier's accuracy evaluation for the Iris dataset. The class is called `MlpLayersTest` and can be found in the `mlp_layers_test.py` file, which is located at the following link:

`https://github.com/PacktPublishing/Hands-On-Genetic-Algorithms-with-Python-Second-Edition/blob/main/chapter_09/mlp_layers_test.py`

The main functionality of this class is highlighted as follows:

1. The `convertParam()` method of the class takes a list called `params`. This is actually the chromosome that we described in the previous subsection and contains the float values that represent up to four hidden layers. The method transforms this list of floats into the `hidden_layer_sizes` tuple:

    ```
    if round(params[1]) <= 0:
        hiddenLayerSizes = round(params[0]),
    elif round(params[2]) <= 0:
        hiddenLayerSizes = (round(params[0]),
                            round(params[1]))
    elif round(params[3]) <= 0:
        hiddenLayerSizes = (round(params[0]),
                            round(params[1]),
                            round(params[2]))
    else:
        hiddenLayerSizes = (round(params[0]),
                            round(params[1]),
                            round(params[2]),
                            round(params[3]))
    ```

2. The `getAccuracy()` method takes the `params` list representing the configuration of the hidden layers, uses the `convertParam()` method to transform it into a `hidden_layer_sizes` tuple, and initializes an MLP classifier with this tuple:

    ```
    hiddenLayerSizes = self.convertParams(params)
    self.classifier = MLPClassifier(
        hidden_layer_sizes=hiddenLayerSizes)
    ```

    Then, it finds the accuracy of the classifier using the same *k-fold cross-validation* calculation that we created for the *Wine dataset* in *Chapter 8, Hyperparameter Tuning of Machine Learning Models*:

    ```
    cv_results = model_selection.cross_val_score(self.classifier,
                                    self.X,
                                    self.y,
                                    cv=self.kfold,
                                    scoring='accuracy')
    return cv_results.mean()
    ```

The `MlpLayersTest` class is utilized by the genetic algorithm-based optimizer. We will explain this part in the next section.

Optimizing the MLP architecture using genetic algorithms

Now that we have a way to represent the architecture configuration of the MLP that's used to classify the Iris flower dataset and a way to determine the accuracy of the MLP for each configuration, we can move on and create a genetic algorithm-based optimizer to search for the configuration – the number of hidden layers (up to 4, in our case) and the number of nodes in each layer – that will yield the best accuracy. This solution is implemented by the `01_optimize_mlp_layers.py` Python program, which is located at the following link:

`https://github.com/PacktPublishing/Hands-On-Genetic-Algorithms-with-Python-Second-Edition/blob/main/chapter_09/01_optimize_mlp_layers.py`

The following steps describe the main parts of this program:

1. We start by setting the lower and upper boundary for each of the float values representing a hidden layer. The first hidden layer is given the range [5, 15], while the rest of the layers start from increasingly larger negative values, which increases their chances of terminating the layer count:

```
# [hidden_layer_layer_1_size, hidden_layer_2_size
#  hidden_layer_3_size, hidden_layer_4_size]
BOUNDS_LOW =  [ 5,  -5, -10, -20]
BOUNDS_HIGH = [15,  10,  10,  10]
```

2. Then, we create an instance of the `MlpLayersTest` class, which will allow us to test the various combinations of the hidden layers' architecture:

```
test = mlp_layers_test.MlpLayersTest(RANDOM_SEED)
```

3. Since our goal is to maximize the accuracy of the classifier, we define a single objective, maximizing fitness strategy:

```
creator.create("FitnessMax", base.Fitness, weights=(1.0,))
```

4. Now, we employ the same approach we used in the previous chapter—since the solution is represented by a list of float values, each of a different range, we use the following loop to iterate over all pairs of lower-bound, upper-bound values, and for each range, we create a separate `toolbox` operator, `layer_size_attribute`, that will later be used to generate random float values in the appropriate range:

```
for i in range(NUM_OF_PARAMS):
    toolbox.register("layer_size_attribute_" + str(i),
                     random.uniform,
                     BOUNDS_LOW[i],
                     BOUNDS_HIGH[i])
```

5. Then, we create a `layer_size_attributes` tuple, which contains the separate float number generators we just created for each hidden layer:

```
layer_size_attributes = ()
for i in range(NUM_OF_PARAMS):
    layer_size_attributes = layer_size_attributes + \
    (toolbox.__getattribute__("layer_size_attribute_" + \
        str(i)),)
```

6. Now, we can use this `layer_size_attributes` tuple in conjunction with DEAP's built-in `initCycle()` operator to create a new `individualCreator` operator that fills up an individual instance with a combination of randomly generated hidden layer-size values:

```
toolbox.register("individualCreator",
                 tools.initCycle,
                 creator.Individual,
                 layer_size_attributes,
                 n=1)
```

7. Then, we instruct the genetic algorithm to use the `getAccuracy()` method of the `MlpLayersTest` instance for fitness evaluation. As a reminder, the `getAccuracy()` method, which we described in the previous subsection, converts the given individual—a list of four floats—into a tuple of hidden layer sizes. These are used to configure the MLP classifier. Then, we train the classifier and evaluate its accuracy using k-fold cross-validation:

```
def classificationAccuracy(individual):
    return test.getAccuracy(individual),
toolbox.register("evaluate", classificationAccuracy)
```

8. As for the genetic operators, we repeat the configuration from the previous chapter. While for the *selection* operator, we use the usual tournament selection with a tournament size of 2, we choose *crossover* and *mutation* operators that are specialized for bounded float-list chromosomes and provide them with the boundaries we defined for each hidden layer:

```
toolbox.register("select",
                 tools.selTournament,
                 tournsize=2)
toolbox.register("mate",
                 tools.cxSimulatedBinaryBounded,
                 low=BOUNDS_LOW,
                 up=BOUNDS_HIGH,
                 eta=CROWDING_FACTOR)
toolbox.register("mutate",
                 tools.mutPolynomialBounded,
                 low=BOUNDS_LOW,
```

```
                    up=BOUNDS_HIGH,
                    eta=CROWDING_FACTOR,
                    indpb=1.0 / NUM_OF_PARAMS)
```

9. In addition, we continue to use the *elitist* approach, where the **hall-of-fame (HOF)** members—the current best individuals—are always passed untouched to the next generation:

```
population, logbook = elitism.eaSimpleWithElitism(population,
        toolbox,
        cxpb=P_CROSSOVER,
        mutpb=P_MUTATION,
        ngen=MAX_GENERATIONS,
        stats=stats,
        halloffame=hof,
        verbose=True)
```

When running the algorithm for 10 generations with a population size of 20, we get the following outcome:

```
gen nevals max avg
0 20 0.666667 0.416333
1 17 0.693333 0.487
2 15 0.76 0.537333
3 14 0.76 0.550667
4 17 0.76 0.568333
5 17 0.76 0.653667
6 14 0.76 0.589333
7 15 0.76 0.618
8 16 0.866667 0.616667
9 16 0.866667 0.666333
10 16 0.866667 0.722667
- Best solution is: 'hidden_layer_sizes'=(15, 5, 8) , accuracy =
0.8666666666666666
```

The preceding results indicate that, within the ranges we defined, the best combination that was found was of three hidden layers of size 15, 5, and 8, respectively. The classification accuracy that we achieved with these values is about 86.7%.

This accuracy seems to be a reasonable result for the problem at hand. However, there's more we can do to improve it even further.

Combining architecture optimization with hyperparameter tuning

While optimizing the network architecture configuration—the hidden layer parameters—we have been using the default (hyper) parameters of the MLP classifier. However, as we saw in the previous chapter, tuning the various hyperparameters has the potential to increase the classifier's performance. Can we incorporate hyperparameter tuning into our optimization? As you may have guessed, the answer is yes. But first, let's take a look at the hyperparameters we would like to optimize.

The `scikit-learn` implementation of the MLP classifier contains numerous tunable hyperparameters. For our demonstration, we will concentrate on the following hyperparameters:

Name	Type	Description	Default Value
activation	enumerated	Activation function for the hidden layers: `{'identity', 'logistic', 'tanh', 'relu'}`	`'relu'`
solver	enumerated	The solver for weight optimization: `{'lbfgs', 'sgd', 'adam'}`	`'adam'`
alpha	float	Strength of the L2 regularization term	`0.0001`
learning_rate	enumerated	Learning rate schedule for weight updates: {'constant', 'invscaling','adaptive'}	`'constant'`

Table 9.1: MLP hyperparameters

As we saw in the previous chapter, a floating point-based chromosome representation allows us to combine various types of hyperparameters into the genetic algorithm-based optimization process. Since we already used a floating-point-based chromosome to represent the configuration of the hidden layers, we can now incorporate other hyperparameters into the optimization process by augmenting the chromosome accordingly. Let's find out how we can do this.

Solution representation

To the existing four floats, representing our network architecture configuration—

$$[n_1, n_2, n_3, n_4]$$—we can add the following four hyperparameters:

- `activation` can have one of four values: `'tanh'`, `'relu'`, `'logistic'`, or `'identity'`. This can be achieved by representing it as a float number in the range of [0, 3.99]. To transform the float value into one of the aforementioned values, we need to apply the `floor()` function to it, which will yield either 0, 1, 2, or 3. We then replace a value of 0 with `'tanh'`, a value of 1 with `'relu'`, a value of 2 with `'logistic'`, and a value of 3 with `'identity'`.

- `solver` can have one of three values: `'sgd'`, `'adam'`, or `'lbfgs'`. Just as with the activation parameter, it can be represented using a float number in the range of [0, 2.99].

- `alpha` is already a float, so no conversion is needed. It will be bound to the range of [0.0001, 2.0].

- `learning_rate` can have one of three values: `'constant'`, `'invscaling'`, or `'adaptive'`. Once again, we can use a float number in the range of [0, 2.99] to represent its value.

Evaluating the classifier's accuracy

The class that will be used to evaluate the MLP classifier's accuracy for the given combination of hidden layers and hyperparameters is called `MlpHyperparametersTest` and is contained in the `mlp_hyperparameters_test.py` file, which is located at the following link:

`https://github.com/PacktPublishing/Hands-On-Genetic-Algorithms-with-Python-Second-Edition/blob/main/chapter_09/mlp_hyperparameters_test.py`

This class is based on the one we used to optimize the configuration of the hidden layers, `MlpLayersTest`, but with a few modifications. Let's go over these:

1. The `convertParam()` method now handles a `params` list, where the first four entries (`params[0]` through `params[3]`) represent the sizes of the hidden layers, just as before, but in addition, `params[4]` through `params[7]` represent the four hyperparameters we added to the evaluation. Consequently, the method has been augmented with the following lines of code, allowing it to transform the rest of the given parameters (`params[4]` through `params[7]`) into their corresponding values, which can then be fed to the MLP classifier:

    ```
    activation = ['tanh', 'relu', 'logistic', 'identity']
    [floor(params[4])]
    solver = ['sgd', 'adam', 'lbfgs'][floor(params[5])]
    alpha = params[6]
    learning_rate = ['constant', 'invscaling',
        'adaptive'][floor(params[7])]
    ```

2. Similarly, the `getAccuracy()` method now handles the augmented `params` list. It configures the MLP classifier with the converted values of all these parameters rather than just the hidden layer's configuration:

```
hiddenLayerSizes, activation, solver, alpha, learning_rate = \
    self.convertParams(params)
self.classifier = MLPClassifier(
    random_state=self.randomSeed,
    hidden_layer_sizes=hiddenLayerSizes,
    activation=activation,
    solver=solver,
    alpha=alpha,
    learning_rate=learning_rate)
```

This `MlpHyperparametersTest` class is utilized by the genetic algorithm-based optimizer. We will look at this in the next section.

Optimizing the MLP's combined configuration using genetic algorithms

The genetic algorithm-based search for the best combination of hidden layers and hyperparameters is implemented by the `02_ptimize_mlp_hyperparameters.py` Python program, which is located at the following link:

`https://github.com/PacktPublishing/Hands-On-Genetic-Algorithms-with-Python-Second-Edition/blob/main/chapter_09/02_optimize_mlp_hyperparameters.py`

Thanks to the unified floating number representation that's used for all parameters, this program is almost identical to the one we used in the previous section to optimize the network architecture. The main difference is in the definition of the BOUNDS_LOW and BOUNDS_HIGH lists, which contain the ranges of the parameters. To the four ranges we defined previously—one for each hidden layer—we now add another four, representing the additional hyperparameters that we discussed earlier in this section:

```
# 'hidden_layer_sizes': first four values
# 'activation'         : 0..3.99
# 'solver'             : 0..2.99
# 'alpha'              : 0.0001..2.0
# 'learning_rate'      : 0..2.99
BOUNDS_LOW  = [ 5,  -5, -10, -20, 0,      0,     0.0001, 0]
BOUNDS_HIGH = [15,  10,  10,  10, 3.999, 2.999, 2.0,     2.999]
```

And that's all it takes—the program is able to handle the added parameters without any further changes.

Running this program produces the following outcome:

```
gen     nevals  max       avg
0       20      0.94      0.605667
1       15      0.94      0.667
2       16      0.94      0.848667
3       17      0.94      0.935
4       17      0.94      0.908667
5       15      0.94      0.936
6       15      0.94      0.889667
7       16      0.94      0.938333
8       17      0.946667         0.938333
9       13      0.946667         0.938667
10      15      0.946667         0.940667
- Best solution is:
'hidden_layer_sizes'=(7, 4, 6)
'activation'='tanh'
'solver'='lbfgs'
'alpha'=1.2786182334834102
'learning_rate'='constant'
=> accuracy =  0.9466666666666667
```

> **Important note**
>
> Please be aware that, due to variations between operating systems, the results that will be produced when you run this program on your system may be somewhat different from what's being shown here.

The preceding results indicate that, within the ranges we defined, the best combination that we found for the hidden layer configuration and hyperparameters was as follows:

- Three hidden layers of size 7, 4, and 6, respectively.
- An `activation` parameter of the `'tanh'` type—instead of the default `'relu'`
- A `solver` parameter of the `'lbfgs'` type—rather than the default `'adam'`
- An `alpha` value of about `1.279` – considerably larger than the default value of 0.0001
- A `learning_rate` parameter of the `'constant'` type—the same as the default value

This combined optimization resulted in a classification accuracy of about 94.7%—a significant improvement over the previous results, all while using fewer nodes than before.

Summary

In this chapter, you were introduced to the basic concepts of ANNs and DL. After getting acquainted with the Iris dataset and the MLP classifier, you were presented with the notion of network architecture optimization. Next, we demonstrated a genetic algorithm-based optimization of network architecture for the MLP classifier. Finally, we were able to combine network architecture optimization with model hyperparameter tuning using the same genetic algorithms approach, thereby enhancing the performance of the classifier even further.

So far, we have concentrated on **supervised learning** (**SL**). In the next chapter, we will look into applying genetic algorithms to **reinforcement learning** (**RL**), an exciting and fast-developing branch of ML.

Further reading

For more information on the topics that we covered in this chapter, please refer to the following resources:

- *Python Deep Learning—Second Edition, Gianmario Spacagna, Daniel Slater, et al., January 16, 2019*

- *Neural Network Projects with Python, James Loy, February 28, 2019*

- `scikit-learn` MLP classifier:

- `https://scikit-learn.org/stable/modules/generated/sklearn.neural_network.MLPClassifier.html`

- *UCI Machine Learning Repository*: `https://archive.ics.uci.edu/`

10

Reinforcement Learning with Genetic Algorithms

In this chapter, we will demonstrate how genetic algorithms can be applied to **reinforcement learning** – a fast-developing branch of machine learning that is capable of tackling complex tasks. We will do this by solving two benchmark environments from the *Gymnasium* (formerly *OpenAI Gym*) toolkit. We will start by providing an overview of reinforcement learning, followed by a brief introduction to *Gymnasium*, a toolkit that can be used to compare and develop reinforcement learning algorithms, as well as a description of its Python-based interface. Then, we will explore two Gymnasium environments, *MountainCar* and *CartPole*, and develop genetic algorithm-based programs to solve the challenges they present.

In this chapter, we will cover the following topics:

- Understanding the basic concepts of reinforcement learning
- Becoming familiar with the *Gymnasium* project and its shared interface
- Using genetic algorithms to solve the *Gymnasium MountainCar* environment
- Using genetic algorithms, in combination with a neural network, to solve the *Gymnasium CartPole* environment

We will start this chapter by outlining the basic concepts of reinforcement learning. If you are a seasoned data scientist, feel free to skip this introductory section.

Technical requirements

In this chapter, we will use Python 3 with the following supporting libraries:

- `deap`

- `numpy`

- `scikit-learn`

- `gymnasium` – *introduced in this chapter*

- `pygame` – *introduced in this chapter*

> **Important Note**
>
> If you use the `requirements.txt` file we provide (See *Chapter 3*), these libraries are already included in your environment.

The *Gymnasium* environments that will be used in this chapter are *MountainCar-v0* (`https://gymnasium.farama.org/environments/classic_control/mountain_car/`) and *CartPole-v1* (`https://gymnasium.farama.org/environments/classic_control/cart_pole/`).

The programs that will be used in this chapter can be found in this book's GitHub repository at `https://github.com/PacktPublishing/Hands-On-Genetic-Algorithms-with-Python-Second-Edition/tree/main/chapter_10`.

Check out the following video to see the code in action: `https://packt.link/OEBOd`.

Reinforcement learning

In the previous chapters, we covered several topics related to machine learning and focused on **supervised learning** tasks. While supervised learning is immensely important and has a lot of real-life applications, reinforcement learning currently seems to be the most exciting and promising branch of machine learning. The reasons for this excitement include the complex, everyday-life-like tasks that reinforcement learning has the potential to handle. In March 2016, *AlphaGo*, a reinforcement learning-based system specializing in the highly complex game of *Go*, was able to defeat the person considered to be the greatest Go player of the past decade in a competition that was widely covered by the media.

While supervised learning requires **labeled data** for training – in other words, pairs of inputs and matching outputs – reinforcement learning does not present immediate right/wrong feedback; instead, it provides an environment where a longer-term, cumulative **reward** is sought after. This means that, sometimes, an algorithm will need to take a momentary step backward to eventually reach a longer-term goal, as we will demonstrate in the first example of this chapter.

The two main components of reinforcement learning task are the **environment** and the **agent**, as illustrated in the following diagram:

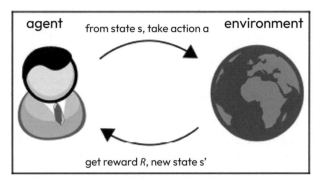

Figure 10.1: Reinforcement learning represented as an interaction
between the agent and the environment

The *agent* represents an algorithm that interacts with the *environment* and attempts to solve a given problem by maximizing the cumulative reward.

The exchange that takes place between the agent and the environment can be expressed as a series of **steps**. In each step, the environment presents the agent with a certain **state** (s), also called an observation. The agent, in turn, performs an **action** (a). The environment responds with a new state (s'), as well as an intermediate reward value (R). This exchange repeats until a certain stopping condition is met. The agent's goal is to maximize the sum of the reward values that are collected along the way.

Despite the simplicity of this formulation, it can be used to describe extremely complex tasks and situations, which makes reinforcement learning applicable to a wide range of applications, such as game theory, healthcare, control systems, supply-chain automation, and operations research.

The versatility of genetic algorithms will be demonstrated once more in this chapter, since we will harness them to assist with reinforcement learning tasks.

Genetic algorithms and reinforcement learning

Various dedicated algorithms have been developed to carry out reinforcement learning tasks – *Q-Learning*, *SARSA*, and *DQN*, to name a few. However, since reinforcement learning tasks involve maximizing a long-term reward, we can think of them as optimization problems. As we have seen throughout this book, genetic algorithms can be used to solve optimization problems of various types. Therefore, genetic algorithms can be utilized for reinforcement learning as well, and in several different ways – two of them will be demonstrated in this chapter. In the first case, our genetic algorithm-based solution will directly provide the agent's optimal series of actions. In the second case, it will supply the optimal parameters for the neural controller that provides these actions.

Before we start applying genetic algorithms to reinforcement learning tasks, let's get acquainted with the toolkit that will be used to conduct these tasks – **Gymnasium**.

Gymnasium

Gymnasium (`https://gymnasium.farama.org/`) – a fork and the official successor of *OpenAI Gym* – is an open source library that was written to allow access to a standardized set of reinforcement learning tasks. It provides a toolkit that can be used to compare and develop reinforcement learning algorithms.

Gymnasium consists of a collection of environments, all presenting a common interface called `env`. This interface decouples the various environments from the agents, which can be implemented in any way we like – the only requirement from the agent is that it can interact with the environment(s) via the `env` interface. This will be described in the next subsection.

The basic package, `gymnasium`, provides access to several environments and can be installed as follows:

```
pip install gymnasium
```

To allow us to render and animate our test environments, the *PyGame* library needs to be installed as well. This can be done using the following command:

```
pip install pygame
```

Several other packages are available, such as "Atari," "Box2D," and "MuJoCo," that provide access to numerous and diverse additional environments. Some of these packages have system dependencies and may only be available for certain operating systems. More information is available at `https://github.com/Farama-Foundation/Gymnasium#installation`.

The next subsection describes interaction with the `env` interface.

The env interface

To create an environment, we need to use the `make()` method and the name of the desired environment, as follows:

```
import gymnasium as gym
env = gym.make('MountainCar-v0')
```

Once the environment has been created, it can be initialized using the `reset()` method, as shown in the following code snippet:

```
observation, info = env.reset()
```

This method returns an `observation` object, describing the initial state of the environment, and a dictionary, `info`, that may contain auxiliary information complementing `observation`. The content of `observation` is environment-dependent.

Conforming with the reinforcement learning cycle that we described in the previous section, the ongoing interaction with the environment consists of sending it an *action* and, in return, receiving an *intermediate reward* and a new *state*. This is implemented by the `step()` method, as follows:

```
observation, reward, terminated, truncated, info = \
    env.step(action)
```

In addition to the `observation` object, which describes the new state and the float `reward` value that represent the interim reward, this method returns the following values:

- `terminated`: A Boolean that turns `true` when the current run (also called *episode*) reaches the terminal state – for example, the agent lost a life, or successfully completed a task.

- `truncated`: A Boolean that can be used to end the episode prematurely before a terminal state is reached – for example, due to a time limit, or if the agent went out of bounds.

- `info`: A dictionary containing optional, additional information that may be useful for debugging. However, it should not be used by the agent for learning.

At any point in time, the environment can be rendered for visual presentation, as follows:

```
env.render()
```

The rendered presentation is environment-specific and is affected by the value of `render_mode`, which can be set when the environment is created. A value of `"human"`, for example, will result in the environment being continuously rendered in the current display or terminal, while the default value of `None` will result in no rendering.

Finally, an environment can be closed to invoke any necessary cleanup, as follows:

```
env.close()
```

If this method isn't called, the environment will automatically close itself the next time Python runs its *garbage collection* process (the process of identifying and freeing memory that is no longer in use by the program), or when the program exits.

> **Note**
>
> Detailed information about the env interface is available at
> `https://gymnasium.farama.org/api/env/`.

The complete cycle of interaction with the environment will be demonstrated in the next section, where we'll encounter our first gymnasium challenge – the *MountainCar* environment.

Solving the MountainCar environment

The `MountainCar-v0` environment simulates a car on a one-dimensional track, situated between two hills. The simulation starts with the car placed between the hills, as shown in the following rendered output:

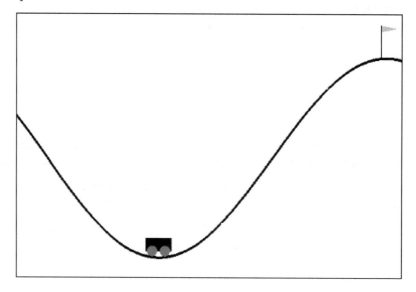

Figure 10.2: The MountainCar simulation – the starting point

The goal is to get the car to climb up the taller hill – the one on the right – and ultimately hit the flag:

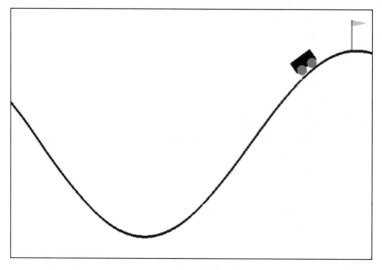

Figure 10.3: The MountainCar simulation – the car climbing the hill on the right

The simulation is set up with the car's engine being too weak to directly climb the taller hill. The only way to reach the goal is to drive the car back and forth until enough momentum is built for climbing. Climbing the left hill can help to achieve this goal, as reaching the left peak will bounce the car back to the right, as shown in the following screenshot:

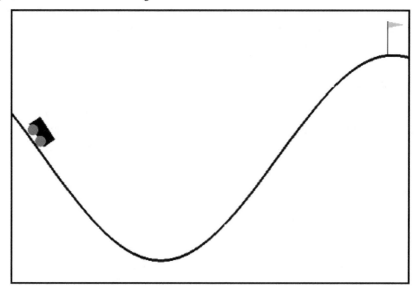

Figure 10.4: The MountainCar simulation – the car bouncing off the hill on the left

This simulation is a great example where intermediate loss (moving left) can help to achieve the ultimate goal (going all the way to the right).

The expected *action* value in this simulation is an integer with of one of the three following values:

- 0: Push left
- 1: No push
- 2: Push right

The `observation` object contains two floats that describe the position and the velocity of the car, such as the following:

```
[-1.0260268, -0.03201975]
```

Finally, the `reward` value is -1 for each time step, until the goal (located at position 0.5) is reached. The simulation will stop after 200 steps if the goal is not reached beforehand.

The goal of this environment is to reach the flag placed on top of the right hill as quickly as possible, and therefore, the agent is penalised with a reward of -1 for each timestep used.

More information about the *MountainCar-v0* environment can be found at

`https://gymnasium.farama.org/environments/classic_control/mountain_car/`.

In our implementation, we will attempt to hit the flag using the least amount of steps, as we apply a sequence of preselected actions from a fixed starting position. To find a sequence of actions that will get the car to climb the tall hill and hit the flag, we will craft a genetic algorithm-based solution. As usual, we will start by defining what a candidate solution for this challenge will look like.

Solution representation

Since *MountainCar* is controlled by a sequence of actions, each with a value of 0 (push left), 1 (no push), or 2 (push right), and there can be up to 200 actions in a single episode, one obvious way to represent a candidate solution is by using a list of length 200, containing values of 0, 1, or 2. An example of this is as follows:

```
[0, 1, 2, 0, 0, 1, 2, 2, 1, ... , 0, 2, 1, 1]
```

The values in the list will be used as actions to control the car and, hopefully, drive it to the flag. If the car made it to the flag in less than 200 steps, the last few items in the list will not be used.

Next, we need to determine how to evaluate a given solution of this form.

Evaluating the solution

While evaluating a given solution, or when comparing two solutions, it is apparent that the reward value alone may not provide us with sufficient information. With the way the reward is defined, its value will always be -200 if we don't hit the flag. When we compare two candidate solutions that don't hit the flag, we would still like to know which one got closer to it and consider it a better solution. Therefore, we will use the final position of the car, in addition to the reward value, to determine the score of the solution:

- If the car did not hit the flag, the score will be the distance from the flag. Therefore, we will look for a solution that minimizes the score.

- If the car hits the flag, the base score will be zero, and from that, we deduct an additional value based on how many unused steps were left, making the score negative. Since we are looking for the lowest score possible, this arrangement will encourage solutions to hit the flag using the smallest possible amount of actions.

This score evaluation procedure is implemented by the `MountainCar` class, which is explored in the following subsection.

The Python problem representation

To encapsulate the MountainCar challenge, we've created a Python class called `MountainCar`. This class is contained in the `mountain_car.py` file, which is located at `https://github.com/PacktPublishing/Hands-On-Genetic-Algorithms-with-Python-Second-Edition/blob/main/chapter_10/mountain_car.py`.

The class is initialized with a random seed and provides the following methods:

- `getScore(actions)`: Calculates the score of a given solution, represented by the list of actions. The score is calculated by initiating an episode of the `MountainCar` environment and running it with the provided actions, and this can be negative if we hit the target with fewer than 200 steps. The lower the score is, the better.
- `saveActions(actions)`: Saves a list of actions to a file using `pickle` (Python's object serialization and deserialization module).
- `replaySavedActions()`: Deserializes the last saved list of actions and replays it using the `replay` method.
- `replay(actions)`: Renders the environment using the "human" `render_mode` and replays the list of actions given to it, visualizing a given solution.

The main method of the class can be used after a solution has been found, serialized, and saved using the `saveActions()` method. The main method will initialize the class and call `replaySavedActions()` to render and animate the last saved solution.

We typically use the main method to animate the best solution that's found by the genetic algorithm-based program. This will be explored in the following subsection.

Genetic algorithms solution

To tackle the *MountainCar* challenge using the genetic algorithms approach, we've created a Python program, `01_solve_mountain_car.py`, which is located at `https://github.com/PacktPublishing/Hands-On-Genetic-Algorithms-with-Python-Second-Edition/blob/main/chapter_10/01_solve_mountain_car.py`.

Since the solution representation we chose for this problem is a list containing the 0, 1, or 2 integer values, this program bears resemblance to the one we used to solve the knapsack 0-1 problem in *Chapter 4, Combinatorial Optimization*, where solutions were represented as lists with the values 0 and 1.

The following steps describe how to create the main parts of this program:

1. We start by creating an instance of the `MountainCar` class, which will allow us to score the various solutions for the *MountainCar* challenge:

```
car = mountain_car.MountainCar(RANDOM_SEED)
```

2.   Since our goal is to minimize the score – in other words, hit the flag with the minimum step count, or get as close as possible to the flag – we define a single objective, minimizing the fitness strategy:

```
creator.create("FitnessMin", base.Fitness, weights=(-1.0,))
```

3.   Now, we need to create a toolbox operator that can produce one of the three allowed action values – 0, 1, or 2:

```
toolbox.register("zeroOneOrTwo", random.randint, 0, 2)
```

4.   This is followed by an operator that fills up an individual instance with these values:

```
toolbox.register("individualCreator",
              tools.initRepeat,
              creator.Individual,
              toolbox.zeroOneOrTwo,
              len(car))
```

5.   Then, we instruct the genetic algorithm to use the `getScore()` method of the `MountainCar` instance for fitness evaluation.

As a reminder, the `getScore()` method, which we described in the previous subsection, initiates an episode of the *MountainCar* environment and uses the given individual – a list of actions – as the inputs to the environment until the episode is done. Then, it evaluates the score – the lower the better – according to the final location of the car. If the car hit the flag, the score can even be negative, based on the number of unused steps left:

```
def carScore(individual):
    return car.getScore(individual),
toolbox.register("evaluate", carScore)
```

6.   As for the genetic operators, we start with the usual *tournament selection*, with a tournament size of 2. Since our solution representation is a list of the 0, 1, or 2 integer values, we can use the *two-point crossover* operator, just like we did when the solution was represented by a list of 0 and 1 values.

For *mutation*, however, rather than the *FlipBit* operator, which is typically used for the binary case, we need to use the *UniformInt* operator, which is used for a range of integer values, and configure it for the range of 0-2:

```
toolbox.register("select", tools.selTournament, tournsize=2)
toolbox.register("mate", tools.cxTwoPoint)
toolbox.register("mutate", tools.mutUniformInt, low=0, up=2,
    indpb=1.0/len(car))
```

7. In addition, we continue to use the *elitist approach*, where the **hall of fame (HOF)** members – the current best individuals – are always passed untouched to the next generation:

```
population, logbook = elitism.eaSimpleWithElitism(population,
    toolbox,
    cxpb=P_CROSSOVER,
    mutpb=P_MUTATION,
    ngen=MAX_GENERATIONS,
    stats=stats,
    halloffame=hof,
    verbose=True)
```

8. After the run, we print the best solution and save it so that we can later animate it, using the replay capability we built into the `MountainCar` class:

```
best = hof.items[0]
print("Best Solution = ", best)
print("Best Fitness = ", best.fitness.values[0])
car.saveActions(best)
```

Running the algorithm for 80 generations and with a population size of 100, we get the following outcome:

```
gen nevals min avg
0       100   0.708709         1.03242
1       78    0.708709         0.975704
...
47      71    0.000170529      0.0300455
48      74    4.87566e-05      0.0207197
49      75    -0.005           0.0150622
50      77    -0.005           0.0121327
...
56      77    -0.02            -0.00321379
57      74    -0.025           -0.00564184
...
79      76    -0.035           -0.0342
80      76    -0.035           -0.03425
Best Solution =   [1, 0, 2, 1, 1, 2, 0, 2, 2, 2, 0, ... , 2, 0,
1, 1, 1, 1, 1, 0]
Best Fitness =   -0.035
```

From the preceding output, we can see that, after about 50 generations, the best solution(s) started hitting the flag, producing score values of zero or less. From hereon, the best solutions hit the flag in fewer steps, thereby producing increasingly negative score values.

As we already mentioned, the best solution was saved at the end of the run, and we can now replay it by running the `mountain_car` program. This replay illustrates how the actions of our solution drive the car back and forth between the two peaks, climbing higher each time, until the car is able to climb the lower hill on the left. Then, it bounces back, which means we have enough momentum to continue climbing the higher hill on the right, ultimately hitting the flag, as shown in the following screenshot:

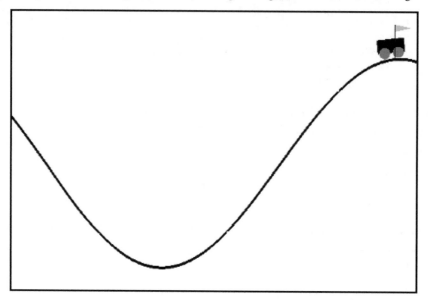

Figure 10.5: The MountainCar simulation – the car reaching the goal

While solving it was a lot of fun, the way this environment is set up did not require us to dynamically interact with it. We were able to climb the hill using a sequence of actions that our algorithm put together, based on the initial location of the car. In contrast, the next environment we are about to tackle – named *CartPole* – requires us to dynamically calculate our action at any time step, based upon the latest observation produced. Read on to find out how we can make this work.

Solving the CartPole environment

The *CartPole-v1* environment simulates a balancing act of a pole, hinged at its bottom to a cart, which moves left and right along a track. Balancing the pole upright is carried out by applying to the cart 1 unit of force – to the right or the left – at a time.

The pole, acting as a pendulum in this environment, starts upright within a small random angle, as shown in the following rendered output:

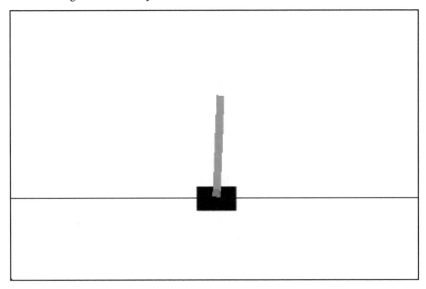

Figure 10.6: The CartPole simulation – the starting point

Our goal is to keep the pendulum from falling over to either side for as long as possible – that is, up to 500 time steps. For every time step that the pole remains upright, we get a reward of +1, so the maximum total reward is 500. The episode will end prematurely if one of the following occurs during the run:

- The angle of the pole from the vertical position exceeds 15 degrees
- The cart's distance from the center exceeds 2.4 units

Consequently, the total reward in these cases will be smaller than 500.

The expected `action` value in this simulation is an integer of one of the two following values:

- 0: Push the cart to the left
- 1: Push the cart to the right

The `observation` object contains four floats that hold the following information:

- **Cart position**, between -2.4 and 2.4
- **Cart velocity**, between -Inf and Inf
- **Pole angle**, between -0.418 rad (-24°) and 0.418 rad (24°)
- **Pole angular velocity**, between -Inf and Inf

For example, we could have an observation of [0.33676587, 0.3786464, -0.00170739, -0.36586074].

More information about the CartPole-v1 environment is available at https://gymnasium.farama.org/environments/classic_control/cart_pole/.

In our proposed solution, we will use these values as inputs at every time step to determine which action to take. We will do this with the aid of a neural network-based controller. This is described in more detail in the following subsection.

Controlling the CartPole with a neural network

To successfully carry out the *CartPole* challenge, we would like to dynamically respond to the changes in the environment. For example, when the pole starts leaning in one direction, we should probably move the cart in that direction but possibly stop pushing when it starts to stabilize. So, the reinforcement learning task here can be thought of as teaching a controller to balance the pole by mapping the four available inputs – cart position, cart velocity, pole angle, and pole velocity – to the appropriate action at each time step. How can we implement such mapping?

One good way to implement this mapping is by using a **neural network**. As we saw in *Chapter 9, Architecture Optimization of Deep Learning Networks*, a neural network, such as a **multilayer perceptron (MLP)**, can implement complex mappings between its inputs and outputs. This mapping is done with the aid of the network parameters – namely, the *weights and biases* of the active nodes in the network, as well as the *transfer functions* that are implemented by these nodes. In our case, we will use a network with a single *hidden layer* of four nodes. In addition, the *input layer* consists of four nodes, one for each of the input values provided by the environment, while the *output layer* has a single node, since we only have one output value – the action to be taken. This network structure can be illustrated with the following diagram:

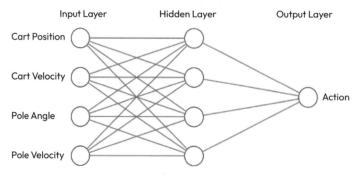

Figure 10.7: The structure of the neural network that's used to control the cart

As we have seen already, the values of the weights and biases of a neural network are typically set during a process in which the network is trained. The interesting part is that, so far, we have only seen this kind of neural network being trained using the backpropagation algorithm while implementing *supervised learning* – that is, in each of the previous cases, we had a training set of inputs and matching outputs, and the network was trained to map each given input to its matching given output. Here, however, as we practice *reinforcement learning*, we don't have such training information available. Instead, we only know how well the network did at the end of the episode. This means that instead of using the conventional training algorithms, we need a method that will allow us to find the best network parameters – weights and biases – based on the results that are obtained by running the environment's episodes. This is exactly the kind of optimization that genetic algorithms are good at – finding a set of parameters that will give us the best results, as long as you have a way to evaluate and compare them. To do that, we need to figure out how to represent the network's parameters, as well as how to evaluate a given set of those parameters. Both of these topics will be covered in the following subsection.

Solution representation and evaluation

Since we have decided to control the cart in the CartPole challenge using a neural network of the *MLP* type, the set of parameters that we will need to optimize are the network's weights and biases, as follows:

- **Input layer**: This layer does not participate in the network mapping; instead, it receives the input values and forwards them to every neuron in the next layer. Therefore, no parameters are needed for this layer.

- **Hidden layer**: Each node in this layer is fully connected to each of the four inputs and, therefore, requires four weights in addition to a single bias value.

- **Output layer**: The single node in this layer is connected to each of the four nodes in the hidden layer and, therefore, requires four weights in addition to a single bias value.

In total, we have 20 weight values and 5 bias values we need to find, all of the `float` type. Therefore, each potential solution can be represented as a list of 25 `float` values, like so:

```
[0.9505049282421143, -0.8068797228337171, -0.45488246459260073, ...
,0.6720551701599038]
```

Evaluating a given solution means creating our MLP with the correct dimensions – four inputs, a four-node hidden layer, and a single output – and assigning the weight and bias values from our list of floats to the various nodes. Then, we need to use this MLP as the controller for the cart pole during one episode. The resulting total reward of the episode is used as the score value for this solution. In contrast to the previous task, here, we aim to *maximize* the score that's achieved. This score evaluation procedure is implemented by the `CartPole` class, which will be explored in the following subsection.

The Python problem representation

To encapsulate the *CartPole* challenge, we've created a Python class called `CartPole`. This class is contained in the `cart_pole.py` file, which is located at `https://github.com/PacktPublishing/Hands-On-Genetic-Algorithms-with-Python-Second-Edition/blob/main/chapter_10/cart_pole.py`.

The class is initialized with an optional random seed and provides the following methods:

- `initMlp()`: Initializes an MLP *regressor* with the desired network architecture (layers) and network parameters (weights and biases), which are derived from the list of floats that represent a candidate solution.

- `getScore()`: Calculates the score of a given solution, represented by the list of float-valued network parameters. This is done by creating a corresponding MLP regressor, initiating an episode of the *CartPole* environment, and running it with the MLP controlling the actions, all while using the observations as inputs. The higher the score is, the better.

- `saveParams()`: Serializes and saves a list of network parameters using `pickle`.

- `replayWithSavedParams()`: Deserializes the latest saved list of network parameters and uses it to replay an episode using the `replay` method.

- `replay()`: Renders the environment and uses the given network parameters to replay an episode, visualizing a given solution.

The main method of the class should be used after a solution has been serialized and saved, using the `saveParams()` method. The main method will initialize the class and call `replayWithSavedParams()` to render and animate the saved solution.

We will typically use the main method to animate the best solution that's found by our genetic algorithm-driven solution, as explored in the following subsection.

A genetic algorithm solution

To interact with the *CartPole* environment and solve it using a genetic algorithm, we've created a Python program, `02_solve_cart-pole.py`, which is located at `https://github.com/PacktPublishing/Hands-On-Genetic-Algorithms-with-Python-Second-Edition/blob/main/chapter_10/02_solve_cart_pole.py`.

Since we will use a list of floats to represent a solution – the network's weights and biases – this program is very similar to the function optimization programs we looked at in *Chapter 6, Optimizing Continuous Functions*, such as the one we used for the *Eggholder function*'s optimization.

The following steps describe how to create the main parts of this program:

1. We start by creating an instance of the `CartPole` class, which will allow us to test the various solutions for the *CartPole* challenge:

   ```
   cartPole = cart_pole.CartPole(RANDOM_SEED)
   ```

2. Next, we set the upper and lower boundaries for the float values we will search for. Since all of our values represent weights and biases within a neural network, this range should be between -1.0 and 1.0 in every dimension:

   ```
   BOUNDS_LOW, BOUNDS_HIGH = -1.0, 1.0
   ```

3. As you may recall, our goal in this challenge is to *maximize* the score – how long we can keep the pole balanced. To do so, we define a single objective, maximizing the fitness strategy:

   ```
   creator.create("FitnessMax", base.Fitness, weights=(1.0,))
   ```

4. Now, we need to create a helper function to create random real numbers that are uniformly distributed within a given range. This function assumes that the range is the same for every dimension, as is the case in our solution:

   ```
   def randomFloat(low, up):
       return [random.uniform(l, u) for l, u in zip([low] * \
           NUM_OF_PARAMS, [up] * NUM_OF_PARAMS)]
   ```

5. Now, we use this function to create an operator that randomly returns a list of floats in the desired range that we set earlier:

   ```
   toolbox.register("attrFloat", randomFloat, BOUNDS_LOW,
       BOUNDS_HIGH)
   ```

6. This is followed by an operator that fills up an individual instance using the preceding operator:

   ```
   toolbox.register("individualCreator",
                    tools.initIterate,
                    creator.Individual,
                    toolbox.attrFloat)
   ```

7. Then, we instruct the genetic algorithm to use the `getScore()` method of the `CartPole` instance for fitness evaluation.

   As a reminder, the `getScore()` method, which we described in the previous subsection, initiates an episode of the *CartPole* environment. During this episode, the cart is controlled by a single-hidden layer MLP. The weight and bias values of this MLP are populated by the list of floats representing the current solution. Throughout the episode, the MLP dynamically maps the observation values of the environment to an action of *right* or *left*. Once the episode is done, the score is set to the total reward, which equates to the number of time steps that the MLP was able to keep the pole balanced – the higher, the better:

```
def score(individual):
    return cartPole.getScore(individual),
toolbox.register("evaluate", score)
```

8. It's time to choose our genetic operators. Once again, we'll use *tournament selection* with a tournament size of 2 as our *selection* operator. Since our solution representation is a list of floats in a given range, we'll use the specialized *continuous bounded crossover* and *mutation* operators provided by the DEAP framework – cxSimulatedBinaryBounded and mutPolynomialBounded, respectively:

```
toolbox.register("select", tools.selTournament, tournsize=2)
toolbox.register("mate",
                 tools.cxSimulatedBinaryBounded,
                 low=BOUNDS_LOW,
                 up=BOUNDS_HIGH,
                 eta=CROWDING_FACTOR)
toolbox.register("mutate",
                 tools.mutPolynomialBounded,
                 low=BOUNDS_LOW,
                 up=BOUNDS_HIGH,
                 eta=CROWDING_FACTOR,
                 indpb=1.0/NUM_OF_PARAMS)
```

9. And, as usual, we use the *elitist approach*, where the HOF members – the current best individuals – are always passed untouched to the next generation:

```
population, logbook = elitism.eaSimpleWithElitism(
    population,
    toolbox,
    cxpb=P_CROSSOVER,
    mutpb=P_MUTATION,
    ngen=MAX_GENERATIONS,
    stats=stats,
    halloffame=hof,
    verbose=True)
```

10. After the run, we print the best solution and save it so that we can animate it, using the replay capability we built into the MountainCar class:

```
best = hof.items[0]
print("Best Solution = ", best)
print("Best Score = ", best.fitness.values[0])
cartPole.saveParams(best)
```

11. In addition, we will run 100 consecutive episodes using our best individual, randomly initiating the CartPole problem each time, so each episode starts from a slightly different starting condition and can potentially yield a different result. We will then calculate the average of all the results:

```
scores = []
for test in range(100):
    scores.append(cart_pole.CartPole().getScore(best))
print("scores = ", scores)
print("Avg. score = ", sum(scores) / len(scores))
```

It is time to find out how well we did in this challenge. By running the algorithm for 10 generations with a population size of 30, we get the following outcome:

```
gen     nevals  max     avg
0       30      68      14.4333
1       26      77      21.7667
...
4       27      381     57.2667
5       26      500     105.733
...
9       22      500     207.133
10      26      500     293.267
Best Solution =  [-0.7441543221198176, 0.34598771744315737,
-0.4221171254602347, ...
Best Score =  500.0
```

From the preceding output, we can see that, after just five generations, the best solution(s) reached the maximum score of 500, balancing the pole for the entire episode's time.

Looking at the results of our additional test, it seems that all 100 tests ended with a perfect score of 500:

```
Running 100 episodes using the best solution...
scores = [500.0, 500.0, 500.0, 500.0, 500.0, 500.0, 500.0, 500.0,
500.0, 500.0, 500.0, 500.0, ... , 500.0]
Avg. score = 500.0
```

As we mentioned previously, each of these 100 runs is done with a slightly different random starting point. However, the controller is powerful enough to balance the pole for the entire run, each and every time. To watch the controller in action, we can play a CartPole episode – or several episodes – with the saved results by launching the cart_pole program. The animation illustrates how the controller dynamically responds to the pole's movement by applying actions that keep the pole balanced on the cart for the entire episode.

If you would like to contrast these results with less-than-perfect ones, you are encouraged to repeat this experiment with three (or even two) nodes in the hidden layer instead of four – just change the `HIDDEN_LAYER` constant value accordingly in the `CartPole` class. Alternatively, you can reduce the number of generations and/or population size of the genetic algorithm.

Summary

In this chapter, you were introduced to the basic concepts of **reinforcement learning**. After getting acquainted with the **Gymnasium** toolkit, you were presented with the *MountainCar* challenge, where a car needs to be controlled in a way that will allow it to climb the taller of two mountains. After solving this challenge using genetic algorithms, you were introduced to the next challenge, *CartPole*, where a cart is to be precisely controlled to keep an upright pole balanced. We were able to solve this challenge by combining the power of a neural network-based controller with genetic algorithm-guided training.

While we have primarily focused on problems involving structured numerical data thus far, the next chapter will shift its focus to applications of genetic algorithms in **Natural Language Processing** (**NLP**), a branch of machine learning that empowers computers to comprehend, interpret, and process human language.

Further reading

For more information, refer to the following resources:

- *Mastering Reinforcement Learning with Python*, Enes Bilgin, December 18 2020
- *Deep Reinforcement Learning Hands-On, 2nd Edition*, Maksim Lapan, January 21, 2020
- Gymnasium documentation:
- `https://gymnasium.farama.org/`
- *OpenAI Gym* (white paper), Greg Brockman, Vicki Cheung, Ludwig Pettersson, Jonas Schneider, John Schulman, Jie Tang, Wojciech Zaremba:
- `https://arxiv.org/abs/1606.01540`

11
Natural Language Processing

This chapter explores how genetic algorithms can enhance the performance of **natural language processing** (**NLP**) tasks while offering insights into their underlying mechanisms.

The chapter begins by introducing the field of NLP and explaining the concept of **word embeddings**. We employ this technique to task a genetic algorithm with playing a *Semantle*-like mystery-word game, challenging it to guess the mystery word.

Subsequently, we investigate **n-grams** and **document classification**. We harness genetic algorithms to pinpoint a compact yet effective subset of features, shedding light on the classifier's operation.

By the end of this chapter, you will have achieved the following:

- Become familiar with the field of NLP and its applications
- Gained an understanding of the concept of word embeddings and their importance
- Implemented a mystery-word game using word embeddings and created a genetic algorithms-driven player to guess the mystery word
- Acquired knowledge about n-grams and their role in document processing
- Developed a process to significantly reduce the size of the feature set used for message classification
- Utilized a minimal feature set to gain insights into the classifier's operation

We will start this chapter with a quick overview of NLP. If you are a seasoned data scientist, feel free to skip the introductory section.

Technical requirements

In this chapter, we will be using Python 3 with the following supporting libraries:

- `deap`
- `numpy`
- `pandas`
- `matplotlib`
- `seaborn`
- `scikit-learn`
- `gensim`—introduced in this chapter

> **Important note**
>
> If you use the `requirements.txt` file we provide (see *Chapter 3*), these libraries are already included in your environment.

The code for this chapter can be found here:

`https://github.com/PacktPublishing/Hands-On-Genetic-Algorithms-with-Python-Second-Edition/tree/main/chapter_11`

Check out the following video to see the code in action:

`https://packt.link/OEBOd`

Understanding NLP

NLP is a fascinating branch of **artificial intelligence** that focuses on the interaction between computers and human language. NLP combines linguistics, computer science, and **machine learning** to enable machines to understand, interpret, and generate human language in a way that's both meaningful and useful. Over the last several years, NLP has been progressively taking on an increasing role in our daily lives, with applications spanning numerous domains, from virtual assistants and chatbots to sentiment analysis, language translation, and information retrieval.

One of the primary goals of NLP is to bridge the communication gap between humans and machines; this is crucial as language is the principal medium through which people interact and communicate their thoughts, ideas, and desires. This goal of bridging the communication gap between humans and machines has driven significant advancements in the field of NLP. A recent notable milestone in this journey is the development of **large language models** (**LLMs**), such as OpenAI's **ChatGPT**.

To create a bridge for human-computer communication, there must be a method in place that can transform human language into numerical representations, allowing machines to understand and process text data more effectively. One such technique is the use of **word embeddings**, described in the next section.

Word embeddings

Word embeddings are numerical representations of words in the English language (or other languages). Each word is encoded using a fixed-length vector of real numbers. These vectors effectively capture semantic and contextual information associated with the words they represent.

Word embeddings are created by training **neural networks** (**NNs**) to create numerical representations for words from large collections of written or spoken texts, where words with similar contexts are mapped to nearby points in a continuous vector space.

Common techniques for creating word embeddings include **Word2Vec**, **Global Vectors for Word Representation** (**GloVe**), and **fastText**.

The typical dimensionality of word embeddings can vary, but common choices are 50, 100, 200, or 300 dimensions. Higher-dimensional embeddings can capture more nuanced relationships but may require more data and computational resources.

For example, the word "dog" in a 50-dimensional Word2Vec embedding space might look something like the following:

```
[0.11008 -0.38781 -0.57615 -0.27714 0.70521 0.53994 -1.0786 -0.40146
1.1504 -0.5678 0.0038977 0.52878 0.64561 0.47262  0.48549 -0.18407
0.1801 0.91397 -1.1979 -0.5778 -0.37985  0.33606 0.772 0.75555 0.45506
-1.7671 -1.0503 0.42566 0.41893 -0.68327 1.5673 0.27685 -0.61708
0.64638 -0.076996 0.37118 0.1308 -0.45137 0.25398 -0.74392 -0.086199
0.24068 -0.64819 0.83549 1.2502 -0.51379 0.04224 -0.88118 0.7158
0.38519]
```

Each of these 50 values represents a different aspect of the word "dog" in the context of the training data. Related words, such as "cat" or "pet," would have word vectors that are closer to the "dog" vector in this space, indicating their semantic similarity. These embeddings not only capture semantic information but also maintain relationships between words, enabling NLP models to understand word relationships, context, and even sentence- and document-level semantics.

The following figure is a 2-dimensional visualization of 50-dimensional vectors representing various English words. This image was created using **t-Distributed Stochastic Neighbor Embedding** (**t-SNE**), a dimensionality reduction technique often used to visualize and explore word embeddings. t-SNE projects word embeddings into a lower-dimensional space while preserving relationships and similarities between data points. This figure demonstrates how certain groups of words, such as fruit or animals, are closer together. Relations between words are apparent as well—for example, the relation between "son" and "boy" resembles that between "daughter" and "girl:"

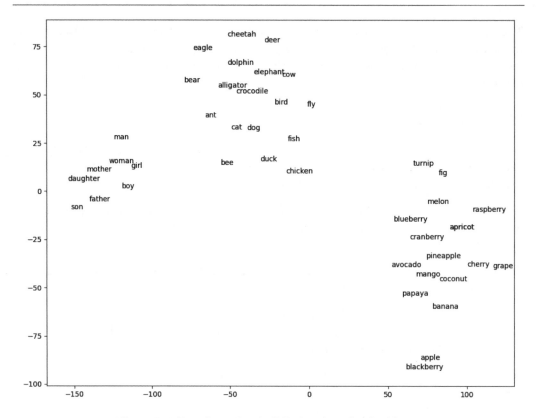

Figure 11.1: Two-dimensional t-SNE plot of word embeddings

In addition to their traditional role in NLP, word embeddings can find use in genetic algorithms, as we will see in the next section.

Word embeddings and genetic algorithms

In previous chapters of this book, we implemented numerous examples of genetic algorithms using fixed-length real-valued vectors (or lists) as the chromosome representation of candidate solutions. Given that word embeddings enable us to represent words (such as "dog") using fixed-length vectors of real-valued numbers, these vectors can effectively serve as the genetic representation of words in genetic algorithm-based applications.

This means we can leverage genetic algorithms to solve problems in which candidate solutions are words in the English language, utilizing word embeddings as the translation mechanism between words and their genetic representation.

To demonstrate this concept, we will embark on solving a fun word game using a genetic algorithm, as described in the next sections.

Finding the mystery word using genetic algorithms

In recent years, online mystery-word games have gained significant popularity. One standout among them is *Semantle*, a game that challenges you to guess the word of the day based on its meaning.

This game provides feedback on how semantically similar your guesses are to the target word and features a "Hot and Cold" meter that indicates the proximity of your guess to the secret word.

Behind the scenes, Semantle employs word embeddings, specifically Word2Vec, to represent both the mystery word and players' guesses. It calculates the semantic similarity between them by measuring the difference between their representations: the closer the vectors, the greater the resemblance between the words. The similarity score returned by the game ranges from -100 (very different from the answer) to 100 (identical to the answer).

In the following subsections, we will create two Python programs. The first serves as a simulation of the Semantle game, while the other embodies a player or solver driven by a genetic algorithm, attempting to uncover the mystery word by maximizing the game's similarity score. Both programs rely on word embedding models; however, to maintain a clear separation, mirroring a real-world scenario, each program employs its own, distinct model. The interaction between the player and the game is limited to exchanging actual guessed words and their corresponding scores, and no embedding vectors are exchanged. This is depicted in the following diagram:

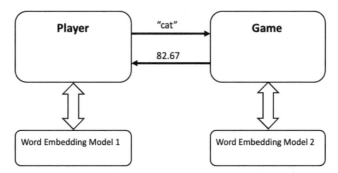

Figure 11.2: Component diagram of the Python modules and their interaction

To add an extra layer of intrigue, we've decided to have each program utilize a completely different embedding model. For that to work, we assume that both embedding models share a substantial overlap in their vocabularies.

The following section provides a detailed account of the Python implementation of these programs.

Python implementation

We will begin by creating the Python implementation of word embedding models using the `gensim` library, as detailed in the following subsection.

The gensim library

The `gensim` library is a versatile Python package primarily recognized for its role in NLP and text analysis tasks. `gensim` simplifies the process of working with word vectors by offering a comprehensive suite of tools for creating, training, and using word embeddings efficiently. One of its key features is its ability to serve as a provider of pre-trained word embedding models, of which we will take advantage in our first Python module, as described next.

The Embeddings class

We start with a Python class called `Embeddings`, encapsulating a `gensim`-based pre-trained word embedding model. This class can be found in the `embeddings.py` file, which is located at the following link:

`https://github.com/PacktPublishing/Hands-On-Genetic-Algorithms-with-Python-Second-Edition/blob/main/chapter_11/embeddings.py`

The main functionality of this class is highlighted as follows:

1. The class's `__init__()` method initializes the random seed (if given), and then proceeds to initialize the chosen (or default) `gensim` model using the `_init_model()` and `_download_and_save_model()` private methods. The former method uploads the model's embedding information from a local file, if available. Otherwise, the latter method downloads the model from the `gensim` repository, separates the essential part for embedding, `KeyedVectors`, and saves it locally to be used the next time:

   ```
   if not isfile(model_path):
       self._download_and_save_model(model_path)
   print(f"Loading model '{self.model_name}' from local file...")
   self.model = KeyedVectors.load_word2vec_format(model_path,
       binary=True)
   ```

2. The `pick_random_embedding()` method can be used to pick a random word out of the model's vocabulary.

3. The `get_similarity()` method is used to retrieve the similarity value of the model between two specified words.

4. The `vec2_nearest_word()` method utilizes the `gensim` model's `similar_by_vector()` method to retrieve the word that is closest to the specified embedding vector. As we will see shortly, this enables the genetic algorithm to use arbitrary vectors (such as randomly generated ones) and have them represent an existing word in the model's vocabulary.

5. Lastly, the `list_models()` method can be used to retrieve and display information about the available embedding models provided by the `gensim` library.

As mentioned earlier, this class is used by both the `Player` and `Game` components, discussed in the following subsections.

The MysteryWordGame class

The `MysteryWordGame` Python class encapsulates the `Game` component. It can be found in the `mystery_word_game.py` file, which is located at the following link:

`https://github.com/PacktPublishing/Hands-On-Genetic-Algorithms-with-Python-Second-Edition/blob/main/chapter_11/mystery_word_game.py`

The main functionality of this class is highlighted as follows:

1. The class employs the `glove-twitter-50` `gensim` pre-trained embedding model developed by Stanford University. This model was specifically designed for Twitter text data and utilizes 50-dimensional embedding vectors.

2. The `__init__()` method of the class initializes the embedding model it will internally use, and then either selects a random mystery word or uses a word specified as an argument for the mystery word:

```
self.embeddings = Embeddings(model_name=MODEL)
    self.mystery_word = given_mystery_word if
        given_mystery_word else
        self.embeddings.pick_random_embedding()
```

3. The `score_guess()` method calculates the score returned by the game for a given guessed word. If the word is not present in the model's vocabulary, which can occur as the player module may use a potentially different model, the score is set to the minimum value of -100. Otherwise, the calculated score value will be a number between -100 and 100:

```
if self.embeddings.has_word(guess_word):
    score = 100 *
    self.embeddings.get_similarity(self.mystery_word,
    guess_word)
else:
    score = -100
```

4. The `main()` method tests the class's functionality by creating an instance of the game, selecting the word `"dog"`, and evaluating several guessed words related to it, such as `"canine"` and `"hound"`. It also includes an unrelated word (`"computer"`) and a word that does not exist in the vocabulary (`"asdghf"`):

```
game = MysteryWordGame(given_mystery_word="dog")
print("-- Checking candidate guess words:")
for guess_word in ["computer", "asdghf", "canine", "hound",
    "poodle", "puppy", "cat", "dog"]:
    score = game.score_guess(guess_word)
    print(f"- current guess: {guess_word.ljust(10)} =>
        score = {score:.2f}")
```

Executing the `main()` method of the class yields the following output:

```
Loading model 'glove-twitter-50' from local file...
--- Mystery word is 'dog' — game on!
-- Checking candidate guess words:
- current guess: computer   => score = 54.05
- current guess: asdghf     => score = -100.00
- current guess: canine     => score = 47.07
- current guess: hound      => score = 64.93
- current guess: poodle     => score = 65.90
- current guess: puppy      => score = 87.90
- current guess: cat        => score = 94.30
- current guess: dog        => score = 100.00
```

We are now ready for the interesting component—the program that attempts to solve the game.

The genetic algorithms-based player program

As mentioned earlier, this module uses a different embedding model from the one used by the game, although it has the option to use the same model. In this case, we have selected the `glove-wiki-gigaword-50` `gensim` pre-trained embedding model, which was trained on a substantial corpus of text from the English *Wikipedia* website and the *Gigaword* dataset.

Solution representation

The solution representation in the genetic algorithm in this case is a real-valued vector (or a list) of the same dimension as the embedding model. This allows each solution to serve as an embedding vector, although not perfectly. Initially, the algorithm employs randomly generated vectors, and through crossover and mutation operations, it's likely that at least some of the vectors won't directly correspond to existing words in the model's vocabulary. To address this issue, we utilize the `vec2_nearest_word()` method from the `Embedding` class, which returns the closest word in the vocabulary. This approach exemplifies the **genotype-to-phenotype mapping** concept, as discussed in *Chapter 4, Combinatorial Optimization*.

Early convergence criteria

In most cases discussed so far, the solution does not possess the knowledge of the best possible score that can be achieved during the optimization process. However, in this case, we know that the best possible score is 100. Once it is achieved, the correct word has been found, and there is no point in continuing the evolutionary cycle. Therefore, we modified the main loop of our genetic algorithm to break if the maximum score is reached. The modified method is called `eaSimple_modified()` and can be found in the `elitism_modified.py` module. It accepts an optional parameter called `max_fitness`. When this parameter is provided with a value, the main loop breaks if the best fitness value found so far reaches or exceeds this value:

```
if max_fitness and halloffame.items[0].fitness.values[0] >=
    max_fitness:
    break
```

Printing out the current best-guessed word

Additionally, the `eaSimple_modified()` method includes the printing of the guessed word corresponding to the individual with the best fitness found so far, as part of the statistics summary generated for every individual:

```
if verbose:
    print(f"{logbook.stream} => {embeddings.vec2_nearest_word(
        np.asarray(halloffame.items[0]))}")
```

The genetic algorithm implementation

The genetic algorithm-based player for the mystery-word game search for the best hyperparameter values is implemented by the `01_find_mystery_word.py` Python program, which is located at the following link:

https://github.com/PacktPublishing/Hands-On-Genetic-Algorithms-with-Python-Second-Edition/blob/main/chapter_11/01_find_mystery_word.py

The following steps describe the main parts of this program:

1. We begin by creating an instance of the `Embeddings` class, which will serve as the word embeddings model for the solver program:

    ```
    embeddings = Embeddings(model_name='glove-wiki-gigaword-50',
        randomSeed=RANDOM_SEED)
    VECTOR_SIZE = embeddings.get_vector_size()
    ```

2. Next, we create an instance of the `MysteryWordGame` class, which represents the game we will be playing. We instruct it to use the word "dog" for demonstration purposes. This word can later be replaced with others, or we can allow the game to choose a random word if we omit the `given_mystery_word` parameter:

    ```
    game = MysteryWordGame(given_mystery_word='dog')
    ```

3. Since our goal is to maximize the game's score, we define a single-objective strategy for maximizing fitness:

```
creator.create("FitnessMax", base.Fitness, weights=(1.0,))
```

4. To create random individuals representing word embeddings, we create a `randomFloat()` function and register it with the toolbox:

```
def randomFloat(low, up):
    return [random.uniform(l, u) for l, u in zip([low] *
        VECTOR_SIZE, [up] * VECTOR_SIZE)]
toolbox.register("attrFloat", randomFloat, BOUNDS_LOW,
    BOUNDS_HIGH)
```

5. The `score()` function is used to evaluate the fitness of each solution, and this process consists of two steps: first, we employ the local `embeddings` model to find the vocabulary word nearest to the evaluated vector (this is where the genotype-to-phenotype mapping takes place). Next, we send this word to the `Game` component and request it to score it as a guessed word. The score returned by the game, a value ranging from -100 to 100, is directly used as the fitness value:

```
def score(individual):
    guess_word = embeddings.vec2_nearest_word(
        np.asarray(individual))
    return game.score_guess(guess_word),
toolbox.register("evaluate", score)
```

6. Now, we need to define genetic operators. While for the *selection* operator, we use the usual *tournament selection* with a tournament size of 2, we choose *crossover* and *mutation* operators that are specialized for bounded float-list chromosomes and provide them with the boundaries we defined for each hyperparameter:

```
toolbox.register("select", tools.selTournament, tournsize=2)
toolbox.register("mate",
                tools.cxSimulatedBinaryBounded,
                low=BOUNDS_LOW,
                up=BOUNDS_HIGH,
                eta=CROWDING_FACTOR)
toolbox.register("mutate",
                tools.mutPolynomialBounded,
                low=BOUNDS_LOW,
                up=BOUNDS_HIGH,
                eta=CROWDING_FACTOR,
                indpb=1.0 / NUM_OF_PARAMS)
```

7. In addition, we continue to employ the elitist approach, where the **Hall of Fame** (HOF) members—the current best individuals—are always passed untouched to the next generation. However, in this iteration, we use the `eaSimple_modified` algorithm, where—in addition—the main loop will terminate when the score reaches the maximum known score:

```
population, logbook = eaSimple_modified(
    population,
    toolbox,
    cxpb=P_CROSSOVER,
    mutpb=P_MUTATION,
    ngen=MAX_GENERATIONS,
    max_fitness=MAX_SCORE,
    stats=stats,
    halloffame=hof,
    verbose=True)
```

By running the algorithm with a population size of 30, we get the following outcome:

```
Loading model 'glove-wiki-gigaword-50' from local file...
Loading model 'glove-twitter-50' from local file...
--- Mistery word is 'dog' — game on!
gen     nevals  max      avg
0       30      51.3262  -43.8478  => stories
1       25      51.3262  -17.5409  => stories
2       26      51.3262  -1.20704  => stories
3       26      51.3262  11.1749   => stories
4       26      64.7724  26.23     => bucket
5       25      64.7724  40.0518   => bucket
6       26      67.487   42.003    => toys
7       26      69.455   37.0863   => family
8       25      69.455   48.1514   => family
9       25      69.455   38.5332   => family
10      27      87.2265  47.9803   => pet
11      26      87.2265  46.3378   => pet
12      27      87.2265  40.0165   => pet
13      27      87.2265  52.6842   => pet
14      26      87.2265  59.186    => pet
15      27      87.2265  41.5553   => pet
16      27      87.2265  49.529    => pet
17      27      87.2265  50.9414   => pet
18      27      87.2265  44.9691   => pet
19      25      87.2265  30.8624   => pet
20      27      100      63.5354   => dog
Best Solution = dog
Best Score = 100.00
```

From this printout, we can observe the following:

- Two distinct word embedding models were loaded, one for the player and the other for the game, as designed

- The mystery word that was set to `dog` was correctly guessed by the genetic algorithm-driven player after 20 generations

- As soon as the word was found, the player quit playing, even though the maximum number of generations was set to 1000

- We can see how the current best-guessed word has evolved:

- `stories` → `bucket` → `toys` → `family` → `pet` → `dog`

This looks great! However, keep in mind that it's just one example. You are encouraged to try out other words, as well as different settings for the genetic algorithm; perhaps even change the embedding models. Are there model pairs that are less compatible than others?

In the next portion of this chapter, we will explore **document classification**.

Document classification

Document classification is a critical task in NLP, involving the categorization of textual documents into predefined classes or categories based on their content. This process is essential for organizing, managing, and extracting meaningful information from large volumes of textual data. Applications of document classification are vast and diverse, spanning various industries and domains.

In the field of information retrieval, document classification plays a crucial role in **search engines**. By categorizing web pages, articles, and documents into relevant topics or genres, search engines can deliver more accurate and targeted search results to users. This enhances the overall user experience and ensures that individuals can quickly access the information they seek.

In customer service and support, document classification enables the **automatic routing** of customer inquiries and messages to the appropriate departments or teams. For instance, emails received by a company can be classified into categories such as "Billing Inquiries," "Technical Support," or "General Inquiries," ensuring that each message reaches the right team for prompt response and resolution.

In the legal domain, document classification is instrumental for tasks such as **e-discovery**, where large volumes of legal documents need to be analyzed for relevance to a case. Classification helps identify documents that are potentially pertinent to a legal matter, streamlining the review process and reducing the time and resources required for legal proceedings.

Moreover, document classification is pivotal in **sentiment analysis**, where it can be used to categorize social media posts, reviews, and comments into positive, negative, or neutral sentiments. This information is invaluable for businesses looking to gauge customer feedback, monitor brand reputation, and make data-driven decisions to improve their products or services.

One effective method for performing document classification is by leveraging n-grams, as elaborated in the upcoming section.

N-grams

An n-gram is a contiguous sequence of *n* items, which can be characters, words, or even phrases, extracted from a larger body of text. By breaking down text into these smaller units, n-grams enable the extraction of valuable linguistic patterns, relationships, and context.

For example, in the case of *character n-grams*, a 3-gram might break the word "apple" into "app," "ppl," and "ple."

Here are some examples of *word n-grams*:

- **Unigrams (1-grams)**:

  Text: "I love to code."

  Unigrams: ["I", "love", "to", "code"]

- **Bigrams (2-grams)**:

  Text: "Natural language processing is fascinating."

  Bigrams: ["Natural language", "language processing", "processing is", "is fascinating"]

- **Trigrams (3-grams)**:

  Text: "Machine learning models can generalize."

  Trigrams: ["Machine learning models", "learning models can", "models can generalize"]

N-grams provide insights into textual content by revealing the sequential arrangement of words or characters, identifying frequent patterns, and extracting features. They help understand language structure, context, and patterns, making them valuable for text analysis tasks such as document classification.

Selecting a subset of n-grams

In *Chapter 7, Enhancing Machine Learning Models Using Feature Selection*, we demonstrated the importance of selecting a meaningful subset of features, known as "feature selection." This process is equally valuable in document classification, especially when dealing with a large number of extracted n-grams, a common occurrence in large documents. The advantages of identifying a relevant subset of n-grams include the following:

- **Dimensionality reduction**: Reducing the number of n-grams makes computations more efficient and prevents overfitting

- **Focus on key features**: Selecting discriminative n-grams helps the model concentrate on crucial features

- **Noise reduction**: Filtering out uninformative n-grams minimizes noise in the data

- **Enhanced generalization**: A well-chosen subset improves the model's ability to handle new documents

- **Efficiency**: Smaller feature sets speed up model training and prediction

Furthermore, identifying a relevant subset of n-grams in document classification can be valuable for model interpretability. By narrowing down the features to a manageable subset, it becomes easier to understand and interpret the factors influencing the model's predictions.

Similarly to what we did in *Chapter 7*, we will apply a genetic algorithms-based search here to identify a relevant subset of n-grams. However, considering that the number of n-grams we anticipate is substantially larger than the number of features in the common datasets we've previously used, we won't be searching for the overall best subset. Instead, our goal will be to find a fixed-size subset of features, such as the best 1,000 or 100 n-grams to use.

Using genetic algorithms to search for a fixed-size subset

As we need to identify a good, fixed-size subset of items within a very large group, let's try to define the usual components needed for the genetic algorithm to work:

- **Solution representation**: Since the subset size is much smaller than the full dataset, it's more efficient to use a fixed-size list of integers representing the indices of the items within the large dataset. For instance, if we aim to create a subset of size 3 from 100 items, a possible solution could be represented as a list, such as [5, 42, 88] or [73, 11, 42].

- **Crossover operation**: To ensure valid offspring, we must prevent the same index from appearing more than once in each offspring. In the previous example, the item "42" appears in both lists. If we used a single-point crossover, for example, we could end up with the offspring [5, 42, 42], which in effect will have only two unique items rather than three. One simple crossover method that overcomes this issue would be as follows:

  I.   Create a set containing all unique items present in both parents. In our example, this set would be {5, 11, 42, 73, 88}.

  II. Generate offspring by randomly selecting from the set mentioned previously. Each offspring should select three items (in this case). A possible result could be [5, 11, 88] and [11, 42, 88].

- **Mutation operation**: A straightforward method to generate a valid mutated individual from an existing one is as follows:

  - For each item in the list, with a specified probability, the item will be replaced by one that does exist in the current list.

  - For example, if we consider the list [11, 42, 88], there's a possibility that the second item (42) could be replaced with, say, 27, resulting in the list [11, 27, 88].

Python implementation

In the following sections, we will implement the following:

- A document classifier that will train on document data from two newsgroups and use n-grams to predict to which newsgroup each document belongs

- A genetic algorithms-driven optimizer that seeks to find the best subset of n-grams to use for this classification task, given the desired size of the subset

We will start with the class implementing the classifier, as described in the next subsection.

Newsgroup document classifier

We start with a Python class called NewsgroupClassifier, implementing a scikit-learn-based document classifier that uses n-grams as features and learns to distinguish between posts from two different newsgroups. This class can be found in the newsgroup_classifier.py file, which is located at the following link:

https://github.com/PacktPublishing/Hands-On-Genetic-Algorithms-with-Python-Second-Edition/blob/main/chapter_11/newsgroup_classifier.py

The main functionality of this class is highlighted as follows:

1. The class's init_data() method, called by __init__(), creates training and testing sets from scikit-learn's built-in dataset of newsgroup posts. It retrieves posts from two categories, 'rec.autos' and 'rec.motorcycles', and preprocesses them to remove headers, footers, and quotes:

```python
categories = ['rec.autos', 'rec.motorcycles']
remove = ('headers', 'footers', 'quotes')
newsgroups_train = fetch_20newsgroups(subset='train',
    categories=categories, remove=remove, shuffle=False)
newsgroups_test = fetch_20newsgroups(subset='test',
    categories=categories, remove=remove, shuffle=False)
```

2. Next, we create two TfidfVectorizer objects: one using word n-grams in the range of 1 to 3 words, and the other using character n-grams in the range of 1 to 10 characters. These vectorizers convert text documents into numerical feature vectors based on the relative frequency of n-grams within each document compared to the entire set of documents. These two vectorizers are then combined into a single vectorizer instance to extract features from the provided newsgroup messages:

```python
word_vectorizer = TfidfVectorizer(analyzer='word',
    sublinear_tf=True, max_df=0.5, min_df=5,
    stop_words="english", ngram_range=(1, 3))
```

```
char_vectorizer = TfidfVectorizer(analyzer='char',
    sublinear_tf=True, max_df=0.5,
    min_df=5, ngram_range=(1, 10))
vectorizer = FeatureUnion([('word_vectorizer', word_vectorizer),
    ('char_vectorizer', char_vectorizer)])
```

3. We proceed by allowing the `vectorizer` instance to "learn" the relevant n-gram information from the training data, and then convert both the training and test data into datasets of vectors containing their corresponding n-gram-based features:

```
self.X_train = vectorizer.fit_transform(newsgroups_train.data)
self.y_train = newsgroups_train.target
self.X_test = vectorizer.transform(newsgroups_test.data)
self.y_test = newsgroups_test.target
```

4. The `get_predictions()` method generates "reduced" versions of both the training and testing sets, utilizing the subset of features provided via the `features_indices` parameter. It subsequently employs an instance of `MultinomialNB`, a classifier commonly used in the context of text classification, which trains on the reduced training set and generates predictions for the reduced testing set:

```
reduced_X_train = self.X_train[:, features_indices]
reduced_X_test = self.X_test[:, features_indices]
classifier = MultinomialNB(alpha=.01)
classifier.fit(reduced_X_train, self.y_train)
return classifier.predict(reduced_X_test)
```

5. The `get_accuracy()` and `get_f1_score()` methods use the `get_predictions()` method to calculate and return the accuracy and the f-score of the classifier, respectively.

6. The `main()` method creates an instance of the `NewsgroupClassifier` class and then trains and tests it using the full set of features. It then repeats the training and evaluation using a random subset of features of the desired size.

    Running the `main()` method yields the following output:

```
Initializing newsgroup data...
Number of features = 51280, train set size = 1192, test set size
= 794
f1 score using all features: 0.8727376310606889
f1 score using random subset of 100 features: 0.589931144127823
```

We can see that using all 51,280 features, the classifier can achieve an f1-score of 0.87, while using a random subset of 100 features has brought the score down to 0.59. Let's find out if selecting the subset of features using a genetic algorithm will enable us to get closer to a higher score.

Finding the best feature subset using a genetic algorithm

The genetic algorithm-based search for the best subset of 100 features (out of the original 51,280) is implemented by the `02_solve_newsgroups.py` Python program, which is located at the following link:

`https://github.com/PacktPublishing/Hands-On-Genetic-Algorithms-with-Python-Second-Edition/blob/main/chapter_11/02_solve_newsgroups.py`

The following steps describe the main parts of this program:

1. We start by creating an instance of the `NewsgroupClassifier` class that will allow us to test the various fixed-size feature subsets:

    ```
    ngc = NewsgroupClassifier(RANDOM_SEED)
    ```

2. We then define two specialized fixed-subset genetic operators, `cxSubset()`—implementing the crossover—and `mutSubset()` — implementing the mutation, as we discussed earlier.

3. Since our goal is to maximize the f1-score of the classifier, we define a single-objective strategy for maximizing fitness:

    ```
    creator.create("FitnessMax", base.Fitness, weights=(1.0,))
    ```

4. To create random individuals representing feature indices, we create a `randomOrder()` function, which utilizes `random.sample()` to generate a random set of indices within the desired range of 51,280. We can then use this function to create individuals:

    ```
    toolbox.register("randomOrder", random.sample, range(len(ngc)),
        SUBSET_SIZE)

    toolbox.register("individualCreator", tools.initIterate,
        creator.Individual, toolbox.randomOrder)
    ```

5. The `get_score()` function is used to evaluate the fitness of each solution (or subset of features) by calling the `get_f1_score()` method of the `NewsgroupClassifier` instance:

    ```
    def get_score(individual):
        return ngc.get_f1_score(individual),
    toolbox.register("evaluate", get_score)
    ```

6. Now, we need to define genetic operators. While for the *selection* operator, we use the usual *tournament selection* with a tournament size of 2, we choose the specialized *crossover* and *mutation* functions that we defined earlier:

    ```
    toolbox.register("select", tools.selTournament, tournsize=2)
    toolbox.register("mate", cxSubset)
    toolbox.register("mutate", mutSubset, indpb=1.0/SUBSET_SIZE)
    ```

7. Finally, it is time to invoke the genetic algorithm flow, where we continue to employ the elitist approach, where the HOF members—the current best individuals—are always passed untouched to the next generation:

```
population, logbook = eaSimple(
    population,
    toolbox,
    cxpb=P_CROSSOVER,
    mutpb=P_MUTATION,
    ngen=MAX_GENERATIONS,
    stats=stats,
    halloffame=hof,
    verbose=True)
```

By running the algorithm for 5 generations with a population size of 30, we get the following outcome:

```
Initializing newsgroup data...
Number of features = 51280, train set size = 1192, test set size = 794
gen     nevals  max                 avg
0       200     0.639922            0.526988
1       166     0.639922            0.544121
2       174     0.663326            0.557525
3       173     0.669138            0.574895
...
198     170     0.852034            0.788416
199     172     0.852034            0.786208
200     167     0.852034            0.788501
-- Best Ever Fitness =  0.8520343720882079
-- Features subset selected =
1:     5074 = char_vectorizer__ crit
2:     12016 = char_vectorizer__=oo
3:     18081 = char_vectorizer__d usi
...
```

The results demonstrate that we successfully identified a subset of 100 features with an f1-score of 85.2%, which is remarkably close to the 87.2% score achieved using all 51,280 features.

When examining the plots displaying the maximum and average fitness over the generations, shown next, it suggests that further improvements might have been possible had we extended the evolutionary process:

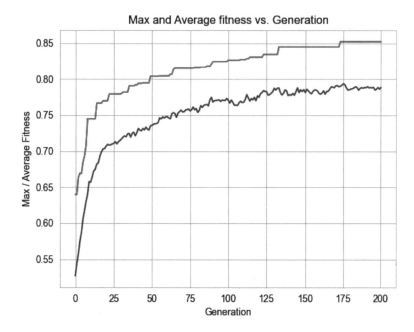

Figure 11.3: Stats of the program searching for the best feature subset

Further reducing the subset size

What if we aim to further reduce the subset size to just 10 features? The outcome may surprise you. By adjusting the SUBSET_SIZE constant to 10, we still achieve a commendable f1-score of 76.1%. Notably, when we examine the 10 selected features, they appear to be fragments of familiar words. In the context of our classification task, which involves distinguishing between posts in a newsgroup dedicated to motorcycles and those related to cars, these features start to reveal their relevance:

```
-- Features subset selected =
 1:    16440 = char_vectorizer__car
 2:    18813 = char_vectorizer__dod
 3:    50905 = char_vectorizer__yamah
 4:    18315 = char_vectorizer__dar
 5:    10373 = char_vectorizer__. The
 6:     6586 = char_vectorizer__ mu
 7:     4747 = char_vectorizer__ bik
 8:     4439 = char_vectorizer__ als
 9:    15260 = char_vectorizer__ave
10:     40719 = char_vectorizer__rcy
```

Removing the character n-grams

The preceding results raise the question of whether we should exclusively utilize word n-grams and eliminate character n-grams. We can implement this by employing a single vectorizer, as follows:

```
vectorizer = TfidfVectorizer(analyzer='word', sublinear_tf=True, max_
df=0.5, min_df=5, stop_words="english", ngram_range=(1, 3))
```

The results of running the `newsgroup_classifier.py` program are as follows:

```
Initializing newsgroup data...
Number of features = 2666, train set size = 1192, test set size = 794
f1 score using all features: 0.8551359241014413
f1 score using random subset of 100 features: 0.6333756056319708
```

These results suggest that exclusively using word n-grams can achieve comparable performance to the original approach while using a significantly smaller feature set (2,666 features).

If we now run the genetic algorithm again, the results are the following:

```
-- Best Ever Fitness =   0.750101164515984
-- Features subset selected =
1:     1669 = oil change
2:     472 = cars
3:     459 = car
4:     361 = bike
5:     725 = detector
6:     303 = autos
7:     296 = auto
8:     998 = ford
9:     2429 = toyota
10:     2510 = v6
```

This set of selected features makes a lot of sense within the context of our classification task and provides insights into how the classifier operates.

Summary

In this chapter, we delved into the rapidly evolving field of NLP. We began by exploring word embeddings and their diverse applications. Our journey led us to experiment with solving the mystery-word game using genetic algorithms, where word embedding vectors served as the genetic chromosome. Following this, we ventured into n-grams and their role in document classification through a newsgroup message classifier. In this context, we harnessed the power of genetic algorithms to identify a compact yet effective subset of n-gram features derived from the dataset.

Finally, we endeavored to minimize the feature subset, aiming to gain insights into the classifier's operations and interpret the factors influencing its predictions. In the next chapter, we will delve deeper into the realm of explainable and interpretable AI while applying genetic algorithms.

Further reading

For more information on the topics that were covered in this chapter, please refer to the following resources:

- *Hands-On Python Natural Language Processing* by *Aman Kedia* and *Mayank Rasu, June 26, 2020*

- *Semantle* word game: `https://semantle.com/`

- `scikit-learn` 20 newsgroups dataset: `https://scikit-learn.org/stable/modules/generated/sklearn.datasets.fetch_20newsgroups.html`

- `scikit-learn TfidfVectorizer`: `https://scikit-learn.org/stable/modules/generated/sklearn.feature_extraction.text.TfidfVectorizer.html`

12

Explainable AI, Causality, and Counterfactuals with Genetic Algorithms

This chapter explores the application of genetic algorithms for generating "what-if" scenarios, providing valuable insights into the analysis of datasets and associated machine learning models, and enabling actionable insights, which help achieve desired outcomes.

This chapter begins by introducing the fields of **Explainable AI (XAI)** and **causality** before explaining the concept of **counterfactuals**. We'll use this technique to explore the ubiquitous *German Credit Risk* dataset and use genetic algorithms to apply a counterfactual analysis to it and discover valuable insights.

By the end of this chapter, you will be able to do the following:

- Be familiar with the fields of XAI and causality, as well as their applications
- Understand the concept of counterfactuals and their importance
- Get acquainted with the German Credit Risk dataset, as well as its shortcomings
- Implement an application to create counterfactual "what-if" scenarios that provide actionable insights for this dataset and shed light on the operation of its associated machine learning model

We will start this chapter with a quick overview of XAI and causality. If you are a seasoned data scientist, feel free to skip this introductory section.

Technical requirements

In this chapter, we will be using Python 3 with the following supporting libraries:

- `deap`
- `numpy`
- `pandas`
- `scikit-learn`

> **Important note**
>
> If you're using the `requirements.txt` file we've provided (see *Chapter 3*), these libraries are already included in your environment.

The programs that will be used in this chapter can be found in this book's GitHub repository at `https://github.com/PacktPublishing/Hands-On-Genetic-Algorithms-with-Python-Second-Edition/tree/main/chapter_12`.

Check out the following video to see the Code in Action:

`https://packt.link/OEBOd`.

Unlocking the black box – XAI

XAI is a crucial element in the realm of **artificial intelligence** (**AI**) that aims to demystify the intricate workings of machine learning models. As AI applications continue to grow, understanding the decision-making processes of models becomes paramount for building trust and ensuring responsible deployment.

XAI intends to address the inherent complexity and opacity of such models and to provide clear and accessible explanations for predictions. This transparency not only enhances the interpretability of AI models but also empowers users, stakeholders, and regulatory bodies to scrutinize and comprehend these processes. In critical domains such as healthcare and finance, where decisions have real-world consequences, XAI is indispensable. For instance, in medical diagnoses, an explainable model not only delivers accurate predictions but also sheds light on specific features in medical images or patient records that influenced the diagnosis, building trust and aligning with ethical standards.

One effective approach to achieving explainability is through *model-agnostic* techniques. These methods offer **post-hoc** ("after the fact") explanations for any machine learning model, regardless of its architecture. Techniques such as *SHAP values* and *LIME* generate explanations by making small, controlled changes to input data or model parameters, revealing the features that contribute most to a prediction.

Building on the foundation of XAI, **causality** adds a layer of depth by exploring the "why" behind the "what," as described in the next section.

Unraveling cause and effect – causality in AI

Knowing not just what AI predicts but also understanding the causal links behind those predictions can be vital, particularly in domains where decisions carry significant consequences.

In AI, causality explores whether changes in aspects of the data impact the predictions or decisions of the model. For example, in healthcare, understanding the causal links between patient parameters and predicted outcomes helps tailor treatments more effectively. The aim is not just accurate predictions but also understanding the mechanisms behind them for a nuanced and actionable insight into the data.

What-if scenarios – counterfactuals

Counterfactuals further augment the interpretability of AI systems by exploring "what-if" scenarios and considering alternative outcomes. Counterfactual explanations help us understand how changes in input could affect model predictions by tweaking these inputs and observing variations (or lack thereof) in the model's decisions. This process essentially poses the question, "*What if?*" and enables us to gain valuable insights into the sensitivity and robustness of AI models.

For example, let's imagine a scenario where an AI-driven car decides to avoid a pedestrian. Through counterfactual analysis, we can uncover how this decision might change under different conditions, providing valuable insights into the model's behavior. As another example, consider a recommendation system. Counterfactuals could help us understand how adjusting certain user preferences might change the recommended items, offering users a clearer understanding of the system's inner workings and allowing developers to enhance user satisfaction.

In addition to providing a deeper understanding of model behavior, counterfactuals can be utilized for model improvement and debugging. By exploring alternative scenarios and observing how changes propagate through the model, developers can identify potential weaknesses, biases, or areas for optimization.

As we will illustrate in the following sections, the exploration of "what-if" scenarios can also empower users to anticipate and interpret the AI system's responses.

Genetic algorithms in counterfactual analysis – navigating alternative scenarios

Genetic algorithms emerge as useful tools for performing counterfactual analysis, offering a flexible approach to modify model inputs to attain a desired outcome. Here, each solution in the genetic algorithm represents a unique input combination. The optimization objective depends on the model's output and can incorporate conditions linked to input values, such as restricting changes or maximizing a particular input value.

In the following sections, we will leverage genetic algorithms to perform counterfactual analysis on a machine learning model tasked with determining the credit risk of a loan applicant. Through this exploration, we aim to answer various questions concerning a particular applicant, gaining insights into the model's inner workings. Additionally, this analysis can provide actionable information to assist the applicants in improving their chances of securing a loan.

Let's begin by familiarizing ourselves with the dataset that will be used to train the model.

The German Credit Risk dataset

For our experiments in this chapter, we will utilize a modified version of the well-known *German Credit Risk* dataset, which is widely used for research and benchmarking in the fields of machine learning and statistics. The original dataset can be accessed from the *UCI Machine Learning Repository* and comprises 1,000 instances, each with 20 attributes. This dataset is designed for a binary classification task that aims to predict whether a loan applicant is creditworthy or poses a credit risk.

In contemporary standards, certain original attributes in this dataset are considered *protected*, notably those representing the gender and age of the candidate. In our adapted version of this dataset, these attributes have been excluded. Additionally, several other features have either been removed or their values have been converted into numerical formats for simplicity.

The modified dataset can be found in the `chapter_12/data/credit_risk_data.csv` file and consists of the following columns:

1. `checking`: The status of the applicant's checking account:

    - 0: No checking account

    - 1: Balance < 100 DM

    - 2: 100 <= balance < 200 DM

    - 3: Balance >= 200 DM

2. `duration`: The duration of the requested loan in months

3. `credit_history`: Information about the applicant's credit history:

    - 0: No loans taken/all loans paid back duly

    - 1: Existing loans paid back duly till now

    - 2: All loans at this bank are paid back duly

    - 3: There's been a delay in paying off in the past

    - 4: Critical account/other loans existing

4. `amount`: The amount of the requested loan

5. `savings`: The status of the applicant's savings account:

    - 0: Unknown/no savings account
    - 1: Balance < 100 DM
    - 2: 100 <= balance < 500 DM
    - 3: 500 <= balance < 1000 DM
    - 4: Balance >= 1000 DM

6. `employment_duration`:

    - 0: Unemployed
    - 1: Duration < 1 year
    - 2: 1 <= duration < 4 years
    - 3: 4 <= duration < 7 years
    - 4: Duration >= 7 years

7. `other_debtors`: Any other individuals who might be co-debtors or share financial responsibility for the loan, aside from the primary applicant:

    - `none`
    - `guarantor`
    - `co-applicant`

8. `present_residence`: The duration of the applicant's current residence at the present address, represented as an integer between 1 and 4

9. `housing`: The housing situation of the applicant:

    - `for free`
    - `own`
    - `rent`

10. `number_credits`: The number of credit accounts held at the same bank

11. `people_liable`: The number of people who are financially dependent on the applicant

12. `telephone`: Whether the applicant has a telephone (1 = yes, 0 = no)

13. `credit_risk`: The value to be predicted:

    - `1`: High risk (indicating a higher likelihood of default or credit issues)

    - `0`: Low risk

For illustration purposes, here are the first 10 lines of the data:

```
1,6,4,1169,0,4,none,4,own,2,1,1,1
2,48,1,5951,1,2,none,2,own,1,1,0,0
0,12,4,2096,1,3,none,3,own,1,2,0,1
1,42,1,7882,1,3,guarantor,4,for free,1,2,0,1
1,24,3,4870,1,2,none,4,for free,2,2,0,0
0,36,1,9055,0,2,none,4,for free,1,2,1,1
0,24,1,2835,3,4,none,4,own,1,1,0,1
2,36,1,6948,1,2,none,2,rent,1,1,1,1
0,12,1,3059,4,3,none,4,own,1,1,0,1
2,30,4,5234,1,0,none,2,own,2,1,0,0
```

While in previous cases of working with datasets, our main goal was to develop a machine learning model for precise predictions on new data, we will now employ counterfactuals to turn things around and identify data that matches a desired prediction.

Exploring counterfactual scenarios for credit risk prediction

As evident from the data, many applicants are deemed a credit risk (indicated by 1 as the last value), leading to loan disapproval. For such applicants, the following question arises: Is there any action they can take to change this decision and be considered creditworthy (indicated by 0 as the outcome)? By *action*, we mean changing the status of one or more of their attributes, such as the amount they are requesting to borrow.

Upon inspecting the dataset, some attributes are challenging or even impossible for an applicant to change, such as employment duration, number of dependents, or current housing. For our examples, we will focus on the following four attributes, all of which offer some flexibility:

- `amount`: The amount of the requested loan

- `duration`: The duration of the requested loan in months

- `checking`: The status of the applicant's checking account

- `savings`: The status of the applicant's savings account

The question can now be phrased as follows: For a candidate currently labeled as a credit risk, what is the minimal change (if any) we could make to these four attributes, or some of them, such that the outcome changes to creditworthy?

To address this question, as well as other relevant questions, we will create the following:

- A machine learning model that has been trained on our dataset. This model will then be used to provide predictions that will enable us to test potential outcomes when an applicant's data is modified.

- A genetic algorithm-based solution that searches for the new attribute values to answer our question(s).

These components will be implemented using Python, as described in the following sections.

The Applicant class

The `Applicant` class represents an applicant from the dataset; in other words, a row from the CSV file. The class also enables us to modify the values of the `amount`, `duration`, `checking`, and `savings` fields, representing the corresponding attributes of the applicant. This class can be found in the `applicant.py` file, which is located at `https://github.com/PacktPublishing/Hands-On-Genetic-Algorithms-with-Python-Second-Edition/blob/main/chapter_12/applicant.py`.

The main functionality of this class is highlighted as follows:

- The class's `__init__()` method uses the value of the `dataset_row` argument to copy the values from the corresponding dataset row and create an instance representing the applicant.

- In addition to setters and getters for the four attributes mentioned previously, the `get_values()` method returns the current values of the four attributes, while the `with_values()` method creates a copy of the original applicant instance and subsequently modifies the copied values of the same four attributes. Both of these methods utilize a list of integers for the four attribute values as they will be used directly by a genetic algorithm that represents potential applicants as a list of four integers.

The CreditRiskData class

The `CreditRiskData` class encapsulates the credit risk dataset and the machine learning model trained on this data. It can be found in the `credit_risk_data.py` file, located at `https://github.com/PacktPublishing/Hands-On-Genetic-Algorithms-with-Python-Second-Edition/blob/main/chapter_12/credit_risk_data.py`.

The main functionality of this class is highlighted in the following steps:

1. The `__init__()` method of the class initializes the random seed; it then calls the `read_dataset()` method, which reads the dataset from the CSV file:

```
self.randomSeed = randomSeed
self.dataset = self.read_dataset()
```

2. Next, it checks if a trained model has already been created and saved to a file. If the model file exists, it is loaded. Otherwise, the `train_model()` method is called.

3. The `train_model()` method created a *Random Forest classifier*, which is first evaluated using a 5-fold cross-validation procedure, to validate its generalization ability:

```
classifier = RandomForestClassifier(
    random_state=self.randomSeed)
kfold = model_selection.KFold(n_splits=NUM_FOLDS)
cv_results = model_selection.cross_val_score(
    classifier, X, y, cv=kfold, scoring='accuracy')
print(f"Model's Mean k-fold accuracy = {cv_results.mean()}")
```

4. Next, the model is trained using the entire dataset and evaluated on it:

```
classifier.fit(X, y)
y_pred = classifier.predict(X)
print(f"Model's Training Accuracy = {accuracy_score(y,
    y_pred)}")
```

5. Once trained, the *Random Forest* model can assign *feature importance* values to the various attributes of the dataset, indicating the contribution of each attribute to the model's predictions. While these values provide insights into the factors influencing the model's decisions, we will use them here to validate our assumption that the four attributes we chose can make a difference:

```
feature_importances = dict(zip(X.columns,
    classifier.feature_importances_))
print(dict(sorted(feature_importances.items(),
    key=lambda item: -item[1])))
```

6. The `is_credit_risk()` method utilizes the trained model to predict the outcome for a given applicant's data using *Scikit-learn*'s `predict()` method, returning `True` when the candidate is considered a credit risk.

7. Additionally, the `risk_probability()` method provides a float between 0 and 1 indicating the degree to which the candidate is considered a credit risk. It utilizes the model's `predict_proba()` method, capturing the continuous output value just before a threshold is applied to convert it into a discrete value of 0 or 1.

8. The convenience method, `get_applicant()`, allows us to select an applicant's row from the dataset and print its data.

9. Finally, the `main()` function starts by creating an instance of the `CreditRiskData` class, which trains the model for the first time if needed. It then retrieves information about the 25th applicant from the dataset and prints it. Afterward, it modifies the values of the four mutable attributes and prints the information of the modified applicant.

10. When executing the `main()` function for the first time, the results of the cross-validation test evaluation, as well as the training accuracy, are printed out:

    ```
    Loading the dataset...
    Model's Mean k-fold accuracy = 0.7620000000000001
    Model's Training Accuracy = 1.0
    ```

    These results indicate that, while the trained model can fully reproduce the dataset's results, its accuracy is approximately 76% when making predictions for unseen samples – a reasonable outcome for this dataset.

11. Next, the feature importance values are printed in descending order. Notably, among the top attributes in the list are the four we chose to modify:

    ```
    ------- Feature Importance values:
    {
        "amount": 0.2357488244229738,
        "duration": 0.15326057481242433,
        "checking": 0.1323559111404014,
        "employment_duration": 0.08332785367394725,
        "credit_history": 0.07824885834794511,
        "savings": 0.06956484835261427,
        "present_residence": 0.06271797270697153,
        ...
    }
    ```

12. The information about the attributes and the predicted outcome for the applicant in row 25 of the dataset is now printed. It's worth noting that in the file, this corresponds to the 27th row, given that the first row contains headers, and the data rows start counting from 0:

    ```
    applicant = credit_data.get_applicant(25)
    ```

    The output is as follows:

    ```
    Before modifications: --------------
    Applicant 25:
    checking                    1
    duration                    6
    credit_history              1
    amount                   1374
    ```

```
savings                          1
employment_duration              2
present_residence                2
...
=> Credit risk = True
```

As the output shows, this applicant is considered a credit risk.

13. The program now modifies all four values using the `with_values()` method:

```
modified_applicant = applicant.with_values([1000, 20, 2, 0])
```

Then, it repeats the printing, reflecting the changes:

```
After modifications: -------------
Applicant 25:
checking                          2
duration                         20
credit_history                    1
amount                         1000
savings                           0
employment_duration               2
present_residence                 2
...
=> Credit risk = False
```

As the preceding output illustrates, when using the new values, the applicant is no longer considered a credit risk. While these modified values were manually selected through trial and error, it's now time to automate this process using genetic algorithms.

Counterfactuals with genetic algorithms

To demonstrate how genetic algorithms can work with counterfactuals, we will start with the same applicant of row 25, who is originally considered a credit risk, and search for the smallest set of changes that will make their prediction creditworthy. As we mentioned previously, we will consider making changes to the `amount`, `duration`, `checking`, and `savings` attributes.

Solution representation

When approaching this problem, a straightforward way to represent a candidate solution is using a list of four integer values, each corresponding to one of the four attributes we want to modify:

[*amount, duration, checking, savings*]

For example, the modified values we used in the main function of the `credit_risk_data.py` program will be represented as follows:

```
[1000, 20, 2, 0]
```

As we did in previous chapters, we will utilize floating-point, real numbers to represent the integers. This enables the genetic algorithm to use the tried-and-true real number operators for crossover and mutation and use the same representation for each item regardless of its range. Before evaluation, the real numbers will be converted into integers using the `int()` function.

We'll evaluate each of these solutions in the next subsection.

Evaluating the solutions

Since our goal is to find the *smallest* degree of changes that will reverse the prediction, a question arises: How do we measure the extent of the changes made? One possible approach is to use the sum of the absolute differences between the current solution's values and the original values, each divided by the range of that value, as shown here:

$$\sum_{i=1}^{4} \frac{|current\ value_i - original\ value_i|}{range\ of\ value_i}$$

Now that we have established the representation and evaluation of the candidate solutions, we are ready to cover the Python implementation of the genetic algorithm.

The genetic algorithm solution

The genetic algorithm-based counterfactual search is implemented in a Python program called `01_counterfactual_search.py`, which is located at `https://github.com/PacktPublishing/Hands-On-Genetic-Algorithms-with-Python-Second-Edition/blob/main/chapter_12/01_counterfactual_search.py`.

The following steps describe the main parts of this program:

1. We begin by defining several constants. Then, we create an instance of the `CreditRiskData` class:

   ```
   credit_data = CreditRiskData(randomSeed=RANDOM_SEED)
   ```

2. The next several code segments are dedicated to setting the ranges of the four attributes that will be used as the solution variables. We start by declaring placeholders, as follows:

   ```
   bounds_low = []
   bounds_high = []
   ranges = []
   ```

   The first list holds the lower bounds of the four attributes, the second holds the upper bounds, and the third holds the difference between them.

3. Now comes the `set_ranges()` method, which accepts the lower and upper bound for each of the four attributes and populates the placeholders accordingly. Since we are using real numbers that will be converted into integers, we'll increment each range to ensure uniform distribution of the resulting integers:

```
bounds_low = [amount_low, duration_low, checking_low,
    savings_low]
bounds_high = [amount_high, duration_high, checking_high,
    savings_high]
bounds_high = [high + 1 for high in bounds_high]
ranges = [high - low for high, low in zip(bounds_high,
    bounds_low)]
```

4. Next, we'll use the `set_ranges()` method to set the ranges for the problem at hand. We've picked the following values:

*   `amount`: 100..5000

*   `duration`: 2..72

*   `checking`: 0..3

*   `savings`: 0..4:

```
bounds_low, bounds_high, ranges =
set_ranges(100, 5000, 2, 72, 0, 3, 0, 4)
```

5. Now, we must select the applicant from row 25 of the dataset (the same one we used earlier) and save its original four values in a separate variable, `applicant_values`:

```
applicant = credit_data.get_applicant(25)
applicant_values = applicant.get_values()
```

6. The `get_score()` function is used to evaluate the fitness of each solution by calculating the cost to be minimized. The cost consists of two parts: first, as mentioned in the *Evaluating the solutions* section, we calculate the distance between the values of the four attributes that the solution represents and the matching original values of the candidate – the larger the distance, the larger the cost:

```
cost = sum(
    [
        abs(int(individual[i]) - applicant_values[i])/ranges[i]
        for i in range(NUM_OF_PARAMS)
    ]
)
```

7. Since we would like the solution to represent a creditworthy candidate, the second, optional, part of the cost is meant to penalize solutions that are considered a credit risk. Here, we'll utilize the `is_credit_risk()` and `risk_probability()` methods so that when the former indicates that the solution is not creditworthy, the latter is used to determine the extent of the added penalty:

```
if credit_data.is_credit_risk(
    applicant.with_values(individual)
):
    cost += PENALTY * credit_data.risk_probability(
        applicant.with_values(individual))
```

8. The rest of the program is very similar to the ones we saw previously when we used a list of real numbers to represent the individuals – for example, the one in *Chapter 9, Architecture Optimization of Deep Learning Networks*. We'll get underway with the single-objective strategy for minimizing fitness as our goal is to minimize the value calculated by the cost function defined earlier:

```
creator.create("FitnessMin", base.Fitness, weights=(-1.0,))
```

9. Since the solution is represented by a list of four float values, one for each of the attributes we can modify, each with its own range, we must define separate toolbox *creator* operators for them using the corresponding `bounds_low` and `bounds_high` values:

```
toolbox.register("amount", random.uniform, \
    bounds_low[0], bounds_high[0])
toolbox.register("duration", random.uniform, \
    bounds_low[1], bounds_high[1])
toolbox.register("checking", random.uniform, \
    bounds_low[2], bounds_high[2])
toolbox.register("savings", random.uniform, \
    bounds_low[3], bounds_high[3])
```

10. These four operators are then used in the definition of `individualCreator`:

```
toolbox.register("individualCreator",
    tools.initCycle,
    creator.Individual,
    (toolbox.amount, toolbox.duration,
        toolbox.checking, toolbox.savings),
    n=1)
```

11. After assigning the *selection* operator to the usual *tournament selection* with a tournament size of 2, we assign the *crossover* and *mutation* operators, which are specialized for bounded float-list chromosomes, and provide them with the ranges we defined earlier:

```
toolbox.register("select",
                 tools.selTournament,
                 tournsize=2)
toolbox.register("mate",
                 tools.cxSimulatedBinaryBounded,
                 low=bounds_low,
                 up= bounds_high,
                 eta=CROWDING_FACTOR)
toolbox.register("mutate",
                 tools.mutPolynomialBounded,
                 low= bounds_low,
                 up=bounds_high,
                 eta=CROWDING_FACTOR,
                 indpb=1.0 / NUM_OF_PARAMS)
```

12. In addition, we'll continue to use the *elitist* approach, where the **hall-of-fame** (**HOF**) members – the current best individuals – are always passed untouched to the next generation:

```
population, logbook = elitism.eaSimpleWithElitism(
    population,
    toolbox,
    cxpb=P_CROSSOVER,
    mutpb=P_MUTATION,
    ngen=MAX_GENERATIONS,
    stats=stats,
    halloffame=hof,
    verbose=True)
```

We end by printing the best solution found, as well as the prediction for that solution.

It's time to try out the program and see the results! The output starts with printing out the original attributes and status of the chosen applicant:

```
Loading the dataset...
Applicant 25:
checking                   1
duration                   6
credit_history             1
amount                  1374
savings                    1
employment_duration        2
```

```
present_residence              2
. . .
=> Credit risk = True
```

As we already know, this candidate is considered a credit risk. Next, the genetic algorithm runs for 30 generations with a population size of 50, attempting to reverse the prediction while making minimal changes to the applicant's four attributes of `amount`, `duration`, `checking`, and `savings`:

```
gen      nevals  min           avg
0        50      0.450063      51.7213
1        42      0.450063      30.2695
2        44      0.393725      14.2223
3        37      0.38311       7.62647
. . .
28       40      0.141661      0.169646
29       40      0.141661      0.175401
30       44      0.141661      0.172197
-- Best solution: Amount = 1370, Duration = 16, checking = 1, savings
= 1
-- Prediction: is_risk = False
```

The preceding output indicates that it is possible to reverse the original prediction and render the applicant creditworthy by extending the duration to 16 months (instead of the original 6). The status of the `checking` and `savings` accounts does not need to be changed.

Although the amount was altered to `1,374`, it could have remained unchanged at 1,374. This can be verified by directly calling the `is_credit_risk()` function with the values [1374, 16, 1, 1]. The way our cost function is defined, the difference between 1,370 and 1,374 divided by the range of 4,900 is minute, potentially causing the algorithm to take many more generations to recognize that the value of 1,374 is better than 1,370. By narrowing the range for the amount to 1,000..2,000, the same program identifies the value of 1,374 well within the allotted 30 generations.

More "what-if" scenarios

We found that altering the loan duration from 6 to 16 months for applicant 25 could render the application creditworthy. However, what if the applicant prefers a shorter duration, or wants to maximize the loan's amount? These are precisely the kind of "what-if" scenarios that counterfactuals explore, and the code we have written enables us to simulate various scenarios and address such questions, as will be demonstrated in the following subsections.

Reducing the duration

Let's start with the case where the same applicant prefers to keep the duration shorter than the value of 16 months found earlier – can other changes compensate for that?

Based on the experience of the previous run, it may prove beneficial to narrow the allowed ranges of the four attributes. Let's try to be more conservative and use the following ranges:

- `amount`: 1000..2000
- `duration`: 2..12
- `checking`: 0..1
- `savings`: 0..1

Here, we'll limit the duration to 12 months and aim to avoid increasing the current balance in either the `checking` or `savings` account. This can be achieved by modifying the call to `set_ranges()`, as follows:

```
bounds_low, bounds_high, ranges = set_ranges(1000, 2000, 2, 12, 0, 1,
0, 1)
```

When we run the modified program, the outcome is the following:

```
-- Best solution: Amount = 1249, Duration = 12, checking = 1, savings
= 1
-- Prediction: is_risk = False
```

This indicates that if the applicant is willing to somewhat lower the requested loan amount, a reduced duration value of 12 months can be achieved.

What if we want to reduce the duration even further? Let's change the duration range to 1..10, for example. This yields the following outcome:

```
-- Best solution: Amount = 1003, Duration = 10, checking = 1, savings
= 0
-- Prediction: is_risk = True
```

This indicates that the algorithm has failed to find a solution where the applicant is creditworthy using these ranges. Note that it does not necessarily mean that such a solution does not exist, but it seems unlikely.

Even if we go back and allow the ranges for the `checking` and `savings` accounts to have their original ranges of 0..3 and 0..4, a solution is not found so long as the duration is limited to 10 months or less. However, allowing the amount to go under 1,000 seems to do the trick. Let's use the following ranges:

- `amount`: 100..2000
- `duration`: 2..10
- `checking`: 0..1
- `savings`: 0..1

Here, we get the following solution:

```
-- Best solution: Amount = 971, Duration = 10, checking = 1, savings =
0
-- Prediction: is_risk = False
```

This implies that if the applicant reduces the loan amount to 971, the application will be approved for the desired duration of 10.

More surprisingly, however, is the value of 0 that was found for the `savings` attribute, which is lower than the original 1. As you may recall, the values of this attribute are interpreted as follows:

- 0: Unknown/no savings account

- 1: Balance < 100 DM

- 2: 100 <= balance < 500 DM

- 3: 500 <= balance < 1000 DM

- 4: Balance >= 1000 DM

It seems unlikely that having no savings would be advantageous when applying for a loan. Furthermore, if we try all possible values other than 0, by setting the range to 1..3, no solution is found. This indicates that, according to the model used, having no savings account is considered preferable to having one, even with a high balance. This is likely an indication of a flaw in the model, or an issue in the dataset itself, such as biased or incomplete data. Such findings are one of the use cases for counterfactuals.

Maximizing the loan's amount

So far, we have manipulated the outcome for an applicant who was initially considered a credit risk. However, we can play this "what-if" game with *any* applicant, including those who were already approved. Let's consider an applicant from row 68 (the 70th line in the file). When printing out the applicant's information, we see the following:

```
Applicant 68:
checking                    0
duration                   36
credit_history              1
amount                   1819
savings                     1
employment_duration         2
present_residence           4
...
=> Credit risk = False
```

Can this applicant be approved for a higher amount without changing the other attributes? To answer this question, we will use a Python program called `02_counterfactual_search.py`, which is located at `https://github.com/PacktPublishing/Hands-On-Genetic-Algorithms-with-Python-Second-Edition/blob/main/chapter_12/02_counterfactual_search.py`.

This program is identical to the previous one, except for three small changes. The first change is the use of this particular applicant:

```
applicant = credit_data.get_applicant(68)
```

The second change is to the range values:

```
bounds_low, bounds_high, ranges = set_ranges(2000, 50000, 36, 36, 0,
    0, 1, 1)
```

The amount range is modified to allow up to a sum of 50,000, while the other ranges have been fixed to the existing values of the candidate. This will enable the genetic algorithm to only modify the amount.

But how do we instruct the genetic algorithm to *maximize* the loan amount? As you may recall, the cost function was initially designed to minimize the distance between the modified individual and the original one within the given range. However, in this scenario, we want the loan amount to be as large as possible compared to the original amount. One approach to address this is to replace the cost function with a new one. However, we'll explore a somewhat simpler solution: we'll set the original loan amount value to the same value we use for the upper end of the range, which is 50,000 in this case. By doing this, when the algorithm aims to find the closest possible solution, it will work inherently to maximize the amount toward this upper limit. This can be done by adding a single line of code that overrides the original amount value of the applicant. The line is placed immediately following the one that stores the original attribute values to be used by the cost function:

```
applicant_values = applicant.get_values()
applicant_values[0] = 50000
```

Note that an index of 0 is used since the `amount` attribute is the first of the four values utilized by the program.

Running this program yields the following output:

```
-- Best solution: Amount = 14165, Duration = 36, checking = 0, savings
= 1
-- Prediction: is_risk = False
```

The preceding output demonstrates that this applicant can maintain the original creditworthy status while increasing the loan amount, without making any changes to the other attributes.

The next question we'll address is whether we can *further* increase the loan amount by allowing changes to the `checking` and/or `savings` attributes. To this effect, we will modify the boundaries so that these two attributes can be adjusted to any valid value:

```
bounds_low, bounds_high, ranges = set_ranges(2000, 50000, 36, 36, 0,
    3, 0, 4)
```

The outcome of the modified program is somewhat surprising:

```
-- Best solution: Amount = 50000, Duration = 36, checking = 1, savings
= 1
-- Prediction: is_risk = False
```

This result suggests that changing the status of the checking account from 0 (no checking account) to 1 (balance < 100 DM) is sufficient to secure a significantly higher loan amount. If we repeat the experiment with much higher amounts (for example, 500,000, replacing the previous value in two different spots in the program), the result will be similar – the candidate is creditworthy for that high amount, so long as the checking status is changed from 0 to 1.

This observation is true when experimenting with other applicants as well, indicating a potential vulnerability in the model.

You are encouraged to make additional modifications to the program and explore your own what-if scenarios. In addition to providing valuable insights and a deeper understanding of the model's behavior, experimenting with "what-if" scenarios can be a lot of fun!

Extending to other datasets

The same process demonstrated in this chapter can be applied to other datasets. For instance, consider a dataset that's used for predicting the expected price of a rental apartment. In this scenario, you can use similar counterfactual analysis to determine what modifications the landlord can introduce to the apartment to achieve a certain rent. Using programs similar to the ones introduced here, you could apply genetic algorithms to explore the sensitivity of the model's predictions to changes in input features and identify actionable insights for achieving desired outcomes.

Summary

In this chapter, you were introduced to the concepts of **XAI**, **causality**, and **counterfactuals**. After getting acquainted with the *German Credis Risk* dataset, we created a machine learning model that predicts whether an applicant is creditworthy. Next, we applied genetic algorithm-based counterfactual analysis of the dataset to the trained model, explored several "what-if" scenarios, and gained valuable insights.

In the next two chapters, we will shift our focus to accelerating the execution of genetic algorithm-based programs, such as the ones we've developed throughout this book, by exploring different strategies for applying concurrency.

Further reading

For more information on the topics that were covered in this chapter, please refer to the following resources:

- *Hands-On Explainable AI (XAI) with Python*, by Denis Rothman, July 2020
- *Causal Inference and Discovery in Python*, by Aleksander Molak, May 2023
- *Responsible AI in the Enterprise*, by Adnan Masood, July 2023
- *Scikit-learn RandomForestClassifier*: `https://scikit-learn.org/stable/modules/generated/sklearn.ensemble.RandomForestClassifier.html`

Part 4:
Enhancing Performance
with Concurrency and
Cloud Strategies

This part focuses on enhancing the performance of genetic algorithms through advanced programming techniques, specifically concurrency and cloud computing. The first chapter introduces concurrency, especially multiprocessing, as a tool to improve genetic algorithm efficiency. The concept is illustrated by applying various multiprocessing methods to a CPU-intensive version of the One-Max problem, demonstrating significant performance gains. Building on this, the next chapter shifts to a client-server model, partitioning the genetic algorithm between asynchronous client operations and server-based fitness function computations. This model is then practically implemented using Flask for the server and Python's asyncio for the client, culminating in deployment to the cloud via AWS Lambda.

This part contains the following chapters:

- *Chapter 13, Accelerating Genetic Algorithms: The Power of Concurrency*
- *Chapter 14, Beyond Local Resources: Scaling Genetic Algorithms in the Cloud*

13

Accelerating Genetic Algorithms – the Power of Concurrency

This chapter delves into the use of concurrency, with a special focus on multiprocessing, as a means to boost the performance of genetic algorithms. We will explore both built-in Python functionalities and an external library to achieve this enhancement.

The chapter starts by highlighting the potential benefits of applying **concurrency** to genetic algorithms. We then proceed to put this theory into practice by experimenting with various **multiprocessing** approaches to a CPU-intensive version of the well-known One-Max problem. This enables us to gauge the extent of performance improvements achievable through these techniques.

By the end of this chapter, you will be able to do the following:

- Understand why genetic algorithms can be computationally intensive and time-consuming

- Recognize why genetic algorithms are well-suited for concurrent execution

- Implement a CPU-intensive version of the One-Max problem, which we have previously explored

- Learn how to use Python's built-in multiprocessing module to accelerate the genetic algorithm process

- Become familiar with the SCOOP library and learn how to integrate it with the DEAP framework to further enhance the efficiency of genetic algorithms

- Experiment with both methods and gain insights into the application of multiprocessing to the problem at hand

Technical requirements

In this chapter, we will use Python 3 with the following supporting libraries:

- `deap`
- `numpy`
- `scoop` – introduced in this chapter

> **Important note**
> If you use the `requirements.txt` file we provide (see *Chapter 3*), these libraries are already included in your environment.

The programs that will be used in this chapter can be found in this book's GitHub repository at `https://github.com/PacktPublishing/Hands-On-Genetic-Algorithms-with-Python-Second-Edition/tree/main/chapter_13`.

Check out the following video to see the Code in Action:

`https://packt.link/OEBOd`

Long runtimes in real-world genetic algorithms

The example programs we've explored so far, while addressing practical problems, were intentionally designed to converge quickly to a reasonable solution. However, in the context of real-world applications, the use of genetic algorithms often proves to be highly time-consuming due to the way they operate – exploring the solution space by considering a diverse set of potential solutions. The main factors affecting the running time of a typical genetic algorithm are as follows:

- **The number of generations**: Genetic algorithms operate through a series of generations, each involving the evaluation, selection, and manipulation of the population.

- **The population size**: Genetic algorithms maintain a population of potential solutions; more complex problems typically require larger populations. This increases the number of individuals that need evaluation, selection, and manipulation in each generation.

- **Fitness evaluation**: The fitness of each individual in the population must be evaluated to determine how well it solves the problem. Depending on the complexity of the fitness function or the nature of the optimization problem, the evaluation process can be both computationally expensive and time-consuming.

- **Genetic operations**: Selection is used to choose pairs of individuals that will serve as parents for each new generation. Crossover and mutation are applied to each of these pairs and, depending on the algorithm's design, can be computationally intensive, especially when dealing with complex data structures. In practice, however, the duration of the fitness function often dominates the time consumed per individual.

One obvious way to mitigate the long running times of genetic algorithms is the use of parallelization, as we will explore further in the following section.

Parallelizing genetic algorithms

Within a single generation, genetic algorithms can be considered **embarrassingly parallelizable** – they can be effortlessly divided into multiple independent tasks, with minimal or no dependency or interaction between them. This is because the fitness evaluation and manipulation of individuals in the population are typically independent tasks. Each individual's fitness is evaluated based on its own characteristics, and genetic operators (crossover and mutation) are applied independently to pairs of individuals. This independence allows for the straightforward parallel execution of these tasks.

Two parallelization methods – **multithreading** and **multiprocessing** – come to mind, as we will explore in the following subsections.

Multithreading

Multithreading is a concurrent execution model that allows multiple threads to exist within the same process, sharing the same resources, such as memory space, but running independently. Each thread represents a separate flow of control, allowing a program to execute multiple tasks concurrently.

In a multithreaded environment, threads can be thought of as lightweight processes that share the same address space. Multithreading is particularly beneficial for tasks that can be divided into smaller, independent units of work, enabling efficient use of available resources and enhancing responsiveness. This is illustrated by the following diagram:

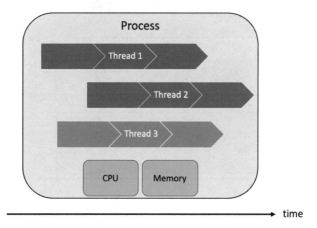

Figure 13.1: Multiple threads running concurrently within a single process

However, multithreading in Python faces some limitations that impact its effectiveness for our use case. A major factor is the **Global Interpreter Lock** (**GIL**) in CPython, the standard implementation of Python. The GIL is a **mutex** (mutually exclusive lock) that protects access to Python objects, preventing multiple native threads from executing Python bytecodes at the same time. As a result, the benefits of multithreading are primarily seen in I/O-bound tasks, as we will explore in the following chapter. For computation-intensive tasks that don't frequently release the GIL, common in many numerical computations, multithreading may not provide the expected performance improvements.

> **Note**
>
> Discussions within the Python community and ongoing research suggest that the limitations imposed by the GIL may be lifted in a future version of Python, potentially enhancing multithreading efficiency.

Fortunately, the approach described next is a highly viable option.

Multiprocessing

Multiprocessing is a concurrent computing paradigm that involves the simultaneous execution of multiple processes within a computer system. In contrast to multithreading, multiprocessing allows for the creation of independent processes, each with its dedicated memory space. These processes can run concurrently on different CPU cores or processors, making it a powerful technique to parallelize tasks and capitalize on modern multi-core architectures, as illustrated in the following diagram:

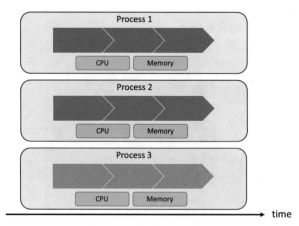

Figure 13.2: Multiple processes running concurrently on separate cores

Operating independently, each process avoids limitations associated with shared memory models, such as the GIL in Python. Multiprocessing proves particularly effective for CPU-bound tasks, commonly encountered in genetic algorithms, where the computational workload can be divided into parallelizable units.

As multiprocessing seems to be a viable way to enhance the performance of genetic algorithms, we will explore its implementation throughout the remainder of this chapter, using a new version of the OneMax problem as our benchmark.

Back to the OneMax problem

In *Chapter 3*, *Using the DEAP Framework*, we utilized the OneMax problem as the "Hello World" of genetic algorithms. As a quick recap, the objective is to discover the binary string of a specified length that maximizes the sum of its digits. For instance, when dealing with a OneMax problem of length 5, candidate solutions such as 10010 (sum of digits = 2) and 01110 (sum of digits = 3) are considered, ultimately leading to the optimal solution of 11111 (sum of digits = 5).

While, in *Chapter 3*, we used a problem length of 100, a population size of 200, and 50 generations, here we will tackle a significantly scaled-down version, having a length of 10, a population size of 20, and only 5 generations. The reasons for this adjustment will become apparent shortly.

A baseline benchmark program

The initial version of this Python program is `01_one_max_start.py`, available at `https://github.com/PacktPublishing/Hands-On-Genetic-Algorithms-with-Python-Second-Edition/blob/main/chapter_13/01_one_max_start.py`.

The main functionality of this program is outlined as follows:

1. Candidate solutions are represented using a list of integers, with values 0 and 1.

2. The `oneMaxFitness()` function calculates the fitness by summing the elements of the list:

    ```
    def oneMaxFitness(individual):
        return sum(individual), # return a tuple
    toolbox.register("evaluate", oneMaxFitness)
    ```

3. For genetic operators, we employ *tournament selection* with a tournament size of 4, *single-point crossover*, and *flip-bit mutation*.

4. The *elitist* approach is applied, utilizing the `elitism.eaSimpleWithElitism()` function.

5. The program's runtime duration is measured using `time.time()` function calls, surrounding the invocation of the `main()` function:

    ```
    if __name__ == "__main__":
        start = time.time()
        main()
        end = time.time()
    print(f"Elapsed time = {(end - start):.2f} seconds")
    ```

Running the program yields the following output:

```
gen     nevals   max      avg
0        20       7        4.35
1        14       7        6.1
2        16       9        6.85
3        16       9        7.6
4        16       9        8.45
5        13       10       8.9
Best Individual =  [1, 1, 1, 1, 1, 1, 1, 1, 1, 1]
Elapsed time = 0.00 seconds
```

The output indicates that the program achieved the optimal solution of 1111111111 by the 5th generation, completing its run in less than 10 milliseconds (considering only the first two decimal digits of the elapsed time).

Another noteworthy detail, which will play a part in subsequent experiments, is the number of fitness evaluations carried out at each generation. The relevant values can be found in the second column from the left, `nevals`. Despite a population size of 20, there are typically fewer than 20 evaluations per generation. This is because the algorithm skips the fitness function if it has already been calculated for a similar individual. Summing the values in this column, we find that the total number of fitness evaluations executed during the program's run amounts to 95.

Simulating computational intensity

As mentioned earlier, the most computationally intensive task in a genetic algorithm is often the fitness evaluation of individuals. To simulate this aspect, we will now intentionally extend the execution time of our fitness function.

This modification is implemented in the Python program, `02_one_max_busy.py`, available at `https://github.com/PacktPublishing/Hands-On-Genetic-Algorithms-with-Python-Second-Edition/blob/main/chapter_13/02_one_max_busy.py`.

This program is based on the previous one, with the following modifications:

1. A `busy_wait()` function is added. This function exercises an empty loop for a specified duration (in seconds):

    ```python
    def busy_wait(duration):
        current_time = time.time()
        while (time.time() < current_time + duration):
            pass
    ```

2. The original fitness function is updated to incorporate a call to the `busy_wait()` function before calculating the sum of the digits:

```
def oneMaxFitness(individual):
    busy_wait(DELAY_SECONDS)
    return sum(individual), # return a tuple
```

3. The `DELAY_SECONDS` constantis added, and its value is set to 3:

```
DELAY_SECONDS = 3
```

Running the modified program yields the following output:

```
gen     nevals  max     avg
0       20      7       4.35
1       14      7       6.1
2       16      9       6.85
3       16      9       7.6
4       16      9       8.45
5       13      10      8.9
Best Individual =   [1, 1, 1, 1, 1, 1, 1, 1, 1, 1]
Elapsed time = 285.01 seconds
```

As anticipated, the output of the modified program is identical to that of the original, with the notable exception of the elapsed time, which has significantly increased to approximately 285 seconds.

This duration makes perfect sense; as highlighted in the previous section, there are 95 executions of the fitness function throughout the program's run (the sum of the values in the `nevals` column). With each execution now taking an additional 3 seconds, the anticipated additional time is calculated as 95 times 3 seconds, totaling 285 seconds.

While examining these results, let's also establish the theoretical limit, or the best-case scenario we can aim for. As indicated in the output, the execution of our benchmark genetic algorithm involves six "rounds" of fitness calculations – one for the initial generation ("generation zero") and one for each of the five subsequent generations. The best achievable time in the context of perfect concurrency within each generation is 3 seconds, equal to the duration of a single fitness evaluation. Therefore, the optimal result we could theoretically achieve would be 18 seconds, calculated as 6 times 3 seconds per round.

With this theoretical limit in mind, we can now proceed to explore the application of multiprocessing to our benchmark.

Multiprocessing using the Pool class

In Python, the `multiprocessing.Pool` module provides a convenient mechanism to parallelize operations across multiple processes. With the `Pool` class, a pool of worker processes is created, allowing tasks to be distributed among them.

The `Pool` class abstracts away the details of managing individual processes by providing the `map` and `apply` methods. Conversely, the DEAP framework makes it very easy to utilize this abstraction. All operations specified in the `toolbox` module are internally executed via a default `map` function. Replacing this map with the `map` from the `Pool` class means that these operations, including fitness evaluations, are now distributed among the worker processes in the pool.

Let's illustrate this by incorporating multiprocessing into our previous program. This modification is implemented in the `03_one_max_pool.py` Python program, available at `https://github.com/PacktPublishing/Hands-On-Genetic-Algorithms-with-Python-Second-Edition/blob/main/chapter_13/03_one_max_pool.py`.

Only a few modifications are required, as outlined here:

1. The `multiprocessing` is imported:

    ```
    import multiprocessing
    ```

2. The `map` method of a `multiprocessing.Pool` class instance is registered as the `map` function to be used by DEAP's toolbox module:

    ```
    toolbox.register("map", pool.map)
    ```

3. The genetic algorithm flow, implemented in the `main()` function, now runs under a `with` statement that manages the creation and cleanup of the `multiprocessing.Pool` instance:

    ```
    def main():
        with multiprocessing.Pool() as pool:
            toolbox.register("map", pool.map)
            # create initial population (generation 0):
            population = toolbox.populationCreator(
                n=POPULATION_SIZE)
            . . .
    ```

Running the modified program on a four-core computer yields the following output:

```
gen     nevals   max    avg
0       20       7      4.35
1       14       7      6.1
2       16       9      6.85
3       16       9      7.6
4       16       9      8.45
5       13       10     8.9
Best Individual =   [1, 1, 1, 1, 1, 1, 1, 1, 1, 1]
Elapsed time = 78.49 seconds
```

As anticipated, the output remains identical to that of the original program, while the duration is significantly shorter compared to the previous one.

> **Important note**
>
> The running time of this program may vary across different computers and even in successive runs on the same machine. As discussed earlier, the optimal result for this benchmark is around 18 seconds. If your computer already approaches this theoretical limit, you can make the benchmark program more CPU-intensive by doubling (or more, if needed) the population size. Remember to adjust all versions of our benchmark program, both in this chapter and the next, to reflect the new population size.

Given the use of a four-core computer, you might expect the duration to be precisely one-quarter of the previous one. However, in this case, we can see that the ratio between the durations is approximately 3.6 ($\approx 285/79$), falling short of the expected 4.

Several factors contribute to us not fully realizing the time-saving potential. A significant factor is the presence of overhead associated with the parallelization process, introducing an additional computational burden when dividing tasks among multiple processes and coordinating their execution.

Moreover, the granularity of the tasks plays a role. While the fitness function consumes most of the processing time, smaller tasks such as the genetic operators of crossover and mutation may encounter a scenario where the overhead of parallelization outweighs the benefits.

Additionally, certain parts of the algorithm, such as handling the hall-of-fame and calculating statistics, are not parallelized. This limitation restricts the extent to which parallelization can be exploited.

To illustrate the last point, let's examine a snapshot of the Activity Monitor application on a Mac while the program is running:

Activity Monitor
All Processes

Process Name	% CPU ⌄
Python	80.4
Python	79.8
Python	78.8
Python	78.5

Figure 13.3: Activity Monitor showing the four genetic algorithm processes in action

As expected, the four Python processes handling the multiprocessor program are heavily utilized, although not at 100%. This prompts the question, can we "squeeze" more out of the CPUs at our disposal and further reduce the duration of the program's run? In the following section, we will explore this possibility.

Increasing the number of processes

Since the four CPUs at our disposal were not utilized at 100%, a question arises: can we further increase utilization by employing more than four concurrent processes?

When we created the instance of the `Pool` class by calling `multiprocessing.Pool()` without any arguments, the number of processes created defaulted to the number of CPUs available – four, in our case. However, we can use the optional `processes` argument to set the desired number of processes, as follows:

```
multiprocessing.Pool(processes=<number of processes>)
```

For our next experiment, we will utilize this option to vary the number of processes and compare the resulting durations. This is implemented in the Python `04_one_max_pool_loop.py` program, available at `https://github.com/PacktPublishing/Hands-On-Genetic-Algorithms-with-Python-Second-Edition/blob/main/chapter_13/04_one_max_pool_loop.py`.

This program introduces a few modifications to the previous one, as outlined here:

1. The `main()` function is renamed `run()`, as we are going to run it several times. It now accepts an argument, `num_processes`.

2. The instantiation of the `Pool` object echoes this argument to create a process pool of the requested size:

```
with multiprocessing.Pool(processes=num_processes) as pool:
```

3. The `plot_graph()` function is added to help illustrate the results.

4. The code launching the program, found at the bottom of the file, now creates a loop, calling the `run()` function multiple times, with the `num_processes` argument incrementing from 1 to 20. Additionally, it collects the resulting durations in a list, `run_times`:

```
run_times = []
for num_processes in range(1, 21):
    start = time.time()
    run(num_processes)
    end = time.time()
    run_time = end - start
    run_times.append(run_time)
```

5. At the end of the loop, the values in the `run_times` list to draw two plots, with the aid of the `plot_graph()` function:

```
plot_graph(1, run_times, "Number of Processes",
    "Run Time (seconds)", hr=33)
plot_graph(2, [1/rt for rt in run_times], "Number of Processes",
    "(1 / Run Time)", "orange")
```

Before we continue to describe the results of this experiment, keep in mind that the actual figures may vary between different computers. As a result, your specific results may differ somewhat. Nevertheless, the main observations should hold true.

Running this program on our four-core computer yields the following output:

```
num_processes = 1
gen      nevals  max      avg
0        20      7        4.35
1        14      7        6.1
2        16      9        6.85
3        16      9        7.6
4        16      9        8.45
5        13      10       8.9
Best Individual =   [1, 1, 1, 1, 1, 1, 1, 1, 1, 1]
Number of Processes = 1 => Run time = 286.62 seconds
num_processes = 2
gen      nevals  max      avg
0        20      7        4.35
1        14      7        6.1
2        16      9        6.85
3        16      9        7.6
4        16      9        8.45
5        13      10       8.9
Best Individual =   [1, 1, 1, 1, 1, 1, 1, 1, 1, 1]
Number of Processes = 2 => Run time = 151.75 seconds
...
num_processes = 20
gen      nevals  max      avg
0        20      7        4.35
1        14      7        6.1
2        16      9        6.85
3        16      9        7.6
4        16      9        8.45
5        13      10       8.9
Best Individual =   [1, 1, 1, 1, 1, 1, 1, 1, 1, 1]
Number of Processes = 20 => Run time = 33.30 seconds
```

In addition, two plots are generated. The first plot, displayed in the following figure, illustrates the runtimes of the program for different numbers of processes. As anticipated, the runtime consistently decreased as the number of processes increased, surpassing the capacity of the four available CPUs. Notably, the improvements became marginal beyond eight processes:

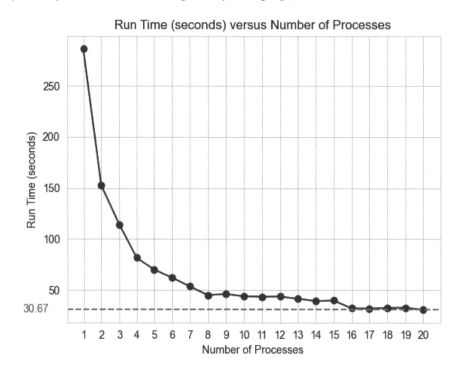

Figure 13.4: The durations of the program run over the different numbers of processes

The dashed red line in this plot represents the shortest duration achieved in our test – about 31 seconds. For the sake of comparison, let's recall the theoretical limit in this test: with 6 rounds of fitness calculations at 3 seconds each, the best possible result is 18 seconds.

The second plot, shown in the following figure, depicts the reciprocal of the duration (or 1/duration), representing the "speed" of the program, across different numbers of processes:

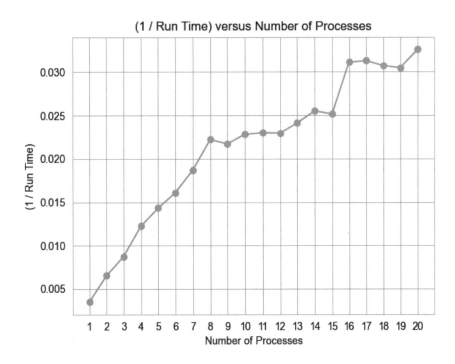

Figure 13.5: The durations of the program run over the different numbers of processes

This plot reveals that the program's speed increases almost linearly with the addition of up to eight processes, but the rate of increase slows beyond this point. Notably, the graph shows a significant performance improvement when moving from 15 to 16 processes, a trend also observable in the previous plot.

The performance gains observed when the number of processes exceeds the number of available physical CPU cores, a phenomenon known as "oversubscription," can be linked to various factors. These include task overlapping, I/O and wait times, multithreading, hyper-threading, and optimizations by the operating system. The marked performance boost from 15 to 16 processes might be influenced by the computer's hardware architecture and the operating system's process scheduling strategies. Additionally, the specific structure of the program's workload, as indicated by the fact that 3 out of 6 rounds of fitness calculations involved exactly 16 fitness evaluations (as shown in the `nevals` column), could also contribute to this increase. It's important to note that these effects can differ, based on the computer's architecture and the nature of the problems being solved.

The takeaway from this experiment is the importance of experimenting with various process counts to find the optimal configuration for your program. Fortunately, you don't need to rerun your entire genetic algorithm each time – a few generations should be enough to compare and figure out the best setup.

Multiprocessing using the SCOOP library

Another approach to introduce multiprocessing is by utilizing **SCOOP**, a Python library designed to parallelize and distribute code execution across multiple processes. **SCOOP**, which stands for **Simple COncurrent Operations in Python**, provides a straightforward interface for parallel computing in Python, which we'll explore shortly.

Applying SCOOP to a DEAP-based program is quite similar to using the `multiprocessing.Pool` module, as demonstrated by the Python `05_one_max_scoop.py` program, available at `https://github.com/PacktPublishing/Hands-On-Genetic-Algorithms-with-Python-Second-Edition/blob/main/chapter_13/05_one_max_scoop.py`.

This program requires only a couple of modifications to the original non-multiprocessing version of the `02_one_max_busy.py` program; these are outlined here:

1. Import SCOOP's `futures` module:

    ```
    from scoop import futures
    ```

2. Register the `map` method of SCOOP's `futures` module as the "map" function to be used by DEAP's toolbox module:

    ```
    toolbox.register("map", futures.map)
    ```

And that's it! However, launching this program requires using SCOOP as the main module, via the following command:

```
python3 -m scoop 05_one_max_scoop.py
```

Running this program on the same four-core computer yields the following output:

```
SCOOP 0.7 2.0 on darwin using Python 3.11.1
Deploying 4 worker(s) over 1 host(s).
Worker distribution:
127.0.0.1: 3 + origin
Launching 4 worker(s) using /bin/zsh.
gen     nevals  max     avg
0       20      7       4.35
1       14      7       6.1
2       16      9       6.85
3       16      9       7.6
4       16      9       8.45
5       13      10      8.9
Best Individual =  [1, 1, 1, 1, 1, 1, 1, 1, 1, 1]
Elapsed time = 78.24 seconds
```

The output starts by providing details about the distribution of the program across four workers; this is followed by the familiar genetic algorithm progress information seen in all the programs so far. Unsurprisingly, the program duration reported in the last line closely resembles the one reported by the `03_one_max_pool.py` program, as both programs employed four concurrent processes.

However, we have seen that "oversubscription" (i.e., using more concurrent processes than the number of available cores) could yield better results. Luckily, SCOOP enables us to control the number of processes, or "workers," via a command-line argument. Let's run the program again but, this time, use 16 workers:

```
python3 -m scoop -n 16 05_one_max_scoop.py
```

The resulting output is as follows:

```
SCOOP 0.7 2.0 on darwin using Python 3.11.1
Deploying 16 worker(s) over 1 host(s).
Worker distribution:
127.0.0.1: 15 + origin
Launching 16 worker(s) using /bin/zsh.
gen      nevals   max       avg
0        20       7         4.35
1        14       7         6.1
2        16       9         6.85
3        16       9         7.6
4        16       9         8.45
5        13       10        8.9
Best Individual =   [1, 1, 1, 1, 1, 1, 1, 1, 1, 1]
Elapsed time = 22.41 seconds
```

The output may sometimes include SCOOP warnings such as `Lost track of future` or `Received an unexpected future`. These warnings indicate communication issues, related to resource constraints caused by oversubscription. Despite these warnings, SCOOP is generally capable of recovering and successfully reproducing the expected results.

A few more experiments reveal that we could achieve times as low as 20 seconds when using SCOOP with process counts of 20 and above. This marks a significant improvement over the 31 seconds we managed with the `multiprocessing.Pool` module for the same problem.

This enhancement might stem from SCOOP's distinct approach to parallelization. For instance, its dynamic task allocation could be more effective than the static method used by `multiprocessing.Pool`. Additionally, SCOOP might handle the overhead of process management more efficiently and could be better at scheduling tasks on the available cores. However, this doesn't mean SCOOP will always have the upper hand over `multiprocessing.Pool`. It's wise to try out both methods and see how they perform with your specific algorithm and problem. The good news is that switching between the two is relatively simple.

Having said that, it's important to mention that SCOOP offers a key feature that sets it apart from `multiprocessing.Pool` – **distributed computing**. This feature allows for parallel processing across multiple machines. We will briefly explore this capability in the following section.

Distributed computing with SCOOP

SCOOP not only supports multiprocessing on a single machine but also enables distributed computing across multiple interconnected nodes. This functionality can be configured in one of two ways:

- **Using the** `--hostfile` **parameter**: This parameter should be followed by the name of a file containing a list of hosts. The format for each line in this file is `<hostname or IP address> <num_of_processes>`, where each line specifies a host and the corresponding number of processes to run on that host.

- **Using the** `--hosts` **parameter**: This option requires a list of hostnames. Each hostname should be listed as many times as the number of processes you intend to run on that host.

For more detailed information and practical examples, you are encouraged to consult SCOOP's official documentation.

A different approach to extending beyond the limitations of a single machine will be explored in the next chapter.

Summary

In this chapter, you were introduced to the concept of applying concurrency to genetic algorithms via multiprocessing, a natural strategy to alleviate their computationally intensive nature. Two main approaches were demonstrated – using Python's built-in `multiprocessing.Pool` class and the SCOOP library. We employed a CPU-intensive version of the familiar One-Max problem as a benchmark, through which several insights were gained. The final part of the chapter addressed the potential of using the SCOOP library for distributed computing.

In the next chapter, we will take the idea of concurrency to the next level by employing a client-server model. This approach will utilize both multiprocessing and multithreading, ultimately leveraging the power of cloud computing to further enhance performance.

Further reading

For more information on the topics that were covered in this chapter, refer to the following resources:

- *Advanced Python Programming: Build high performance, concurrent, and multi-threaded apps with Python using proven design patterns* by Dr. Gabriele Lanaro, Quan Nguyen, and Sakis Kasampalis, February 2019

- SCOOP framework documentation: `https://scoop.readthedocs.io/en/latest/`

- Python multiprocessing module documentation: `https://docs.python.org/3/library/multiprocessing.html`

14

Beyond Local Resources – Scaling Genetic Algorithms in the Cloud

This chapter builds on the previous one, which focused on using multiprocessing to enhance genetic algorithm performance. It restructures the genetic algorithm into a **client-server** model, where the client employs **asynchronous I/O**, and the server manages fitness function calculations. The server component is then deployed to the cloud via **AWS Lambda**, demonstrating a practical application of serverless architecture in optimizing genetic algorithm computations.

This chapter starts by discussing the advantages of dividing genetic algorithms into client and server components. It then progresses to implementing this client-server model while using the same One-Max benchmark problem from the previous chapter. The server is built using Flask, while the client leverages Python's `asyncio` library for asynchronous operations. The chapter includes experiments with deploying the Flask application on production-grade servers before ultimately deploying it on AWS Lambda, a serverless computing service, showcasing how cloud resources can be used to enhance the computational efficiency of genetic algorithms.

In this chapter, you will do the following:

- Understand the restructuring of genetic algorithms into a client-server model
- Learn how to use Flask to create a server that performs fitness calculations
- Develop an asynchronous I/O client in Python that interacts with the Flask server for genetic algorithm evaluations
- Gain familiarity with Python WSGI HTTP servers such as Gunicorn and Waitress
- Learn how to deploy the Flask server component to the cloud using Zappa for serverless execution on AWS Lambda

Technical requirements

In this chapter, we will be using Python 3 with the following supporting libraries:

- `deap`
- `numpy`
- `aiohttp` – introduced in this chapter

> **Important note**
>
> If you use the `requirements.txt` file we provide (see *Chapter 3*), these libraries are already included in your environment.

In addition, a separate server virtual environment will be created and used for a separate server module, with the following supporting libraries:

- `flask`
- `gunicorn`
- `waitress`
- `zappa`

The programs that will be used in this chapter can be found in this book's GitHub repository at `https://github.com/PacktPublishing/Hands-On-Genetic-Algorithms-with-Python-Second-Edition/tree/main/chapter_14`.

Check out the following video to see the Code in Action:

`https://packt.link/OEBOd`

The next level in genetic algorithm performance – embracing a client-server architecture

In the previous chapter, we implemented our genetic algorithm in a multiprocessor approach, leveraging the "embarrassingly parallelizable" nature of genetic algorithms to significantly reduce the time required for each generation to complete. Recognizing that the most time-consuming aspect of the algorithm is typically the computation of the fitness function, we established a benchmark that simulates a CPU-intensive fitness function. By employing Python's built-in multiprocessing capabilities as well as an external library named SCOOP, we successfully managed to decrease the runtime substantially.

However, these implementations were constrained to a single program operating on a single machine. When addressing real-world problems, this approach is likely to encounter resource limitations of the machine—not just in terms of available CPU cores but also in essential resources such as memory and storage. In contrast to our benchmark program, which primarily consumes CPU time, real-world fitness functions may have extensive demands for both processing power and memory, presenting a significant challenge.

We have observed that the SCOOP library supports distributed computing by utilizing additional machines on our network. However, in this chapter, we will explore a different approach that involves dividing our program into two separate components. This methodology will afford us greater flexibility in selecting the platforms for these components. Such a strategy opens up opportunities for more diverse and powerful computing solutions, including cloud-based services or specialized hardware, thereby overcoming some limitations inherent in relying solely on networked machines.

The upcoming sections will detail the design and implementation of this new structure, which we will carry out in several stages.

Implementing a client-server model

Our plan is to divide the execution of the genetic algorithm into two distinct parts—client and server—outlined as follows:

- **Client Component**: The client will oversee the evolutionary logic in a centralized manner. This includes managing the initialization of populations, selection processes, and genetic operations such as crossover and mutation.

- **Server Component**: The server will be tasked with executing the resource-intensive fitness function calculations. It will employ multiprocessing to fully leverage its computational resources, circumventing the limitations imposed by Python's **Global Interpreter Lock (GIL)**.

- **Client's Use of Asynchronous I/O**: Additionally, the client will employ asynchronous I/O, which operates on a single-threaded, event-driven model. This approach can efficiently handle I/O-bound tasks, allowing the program to concurrently manage other operations while waiting for I/O processes to complete. In adopting asynchronous I/O for server communication, the client is able to send a request and then proceed with other tasks, rather than waiting passively for a response. This is akin to a waiter who delivers a guest's order to the kitchen and, instead of waiting there, takes the next order from another table while the previous orders are being prepared. Similarly, our client optimizes the workflow by not blocking the main thread of execution while waiting for server responses.

This client-server model and its operational dynamics are illustrated in the following diagram:

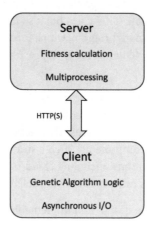

Figure 14.1: Block diagram of the proposed client-server setup

Before we dive into implementing this model, it's highly recommended to set up a separate Python environment specifically for the server component, as described in the next section.

Using a separate environment

As we began coding in *Chapter 3, Using the DEAP Framework*, we recommended creating a virtual environment for our programs, utilizing tools such as venv or conda. Employing a virtual environment is a best practice in Python development, as it allows us to keep the dependencies of our projects isolated from both other Python projects and the system's default settings and dependencies.

Given that the client part manages the logic of the genetic algorithm and uses the DEAP framework, we can continue developing it within the same environment we've been using so far. However, it's advisable to create a separate environment for the server component. The reasoning is twofold: firstly, the server will not be using the DEAP dependency but will instead rely on a different set of Python libraries; and secondly, we ultimately plan to deploy the server outside our local computer, so it's preferable to have this deployment as lightweight as possible.

For reference, you can review the process of creating a virtual environment using venv here: https://docs.python.org/3/library/venv.html.

Similarly, environment management using conda is described here: https://conda.io/projects/conda/en/latest/user-guide/tasks/manage-environments.html.

The code for the server module is best kept in a separate directory as well; in our repository, we keep it in a directory named server, under the chapter_13 directory.

Once we have created and activated a virtual environment for the server component, we're ready to dive into the coding of the components. But first, let's take a moment to recap the benchmark problem we're tackling.

Revisiting the One-Max problem, yet again

As a reminder, in *Chapter 13*, *Accelerating Genetic Algorithms*, we employed a version of the OneMax problem as a benchmark. The goal of this program is to find the binary string of a specified length that maximizes the sum of its digits. For our purposes, we used a reduced problem length of 10 digits, along with smaller figures for the population size and the number of generations. Additionally, we introduced a `busy_wait()` function into the original fitness evaluation function. This function kept the CPU busy for three seconds during each evaluation, significantly increasing the program's execution time. This setup allowed us to experiment with various multiprocessing schemes and to compare their respective running durations.

For our experiments with the client-server model, we will continue to use this same program, albeit with modifications tailored to our current requirements. This approach will allow us to directly compare the results with those obtained in the previous chapter.

We are finally ready to write some code—starting with the server module, which will be based on Flask, the Python web application framework.

Creating the server component

Flask, known for its lightweight and flexible nature, will be the cornerstone of our server component in the Python environment. Its simplicity and user-friendly approach make it a popular choice, especially for projects that require adaptability across various platforms and cloud installations.

To install Flask, make sure you are within the server's virtual environment, and use the following command:

```
pip install Flask
```

One of the key advantages of using Flask is that it requires minimal coding effort on our part. We just need to write the handlers for incoming requests, and Flask efficiently manages all the other underlying processes. Given that our server component's primary responsibility is to handle the calculation of the fitness function, the amount of code we need to write is minimal.

The relevant Python program we created is `fitness_evaluator.py`, available at the following link:

https://github.com/PacktPublishing/Hands-On-Genetic-Algorithms-with-Python-Second-Edition/blob/main/chapter_14/server/fitness_evaluator.py

This program extends the minimal application outlined in Flask's Quickstart documentation, as detailed here:

1. First, we import the `Flask` class and create an instance of this class. The `__name__` argument represents the name of the application's module or package:

```
from flask import Flask
app = Flask(__name__)
```

2. Next, we define the `welcome()` function for "health-check" purposes. This function returns an HTML-formatted welcome message. When we direct our browser to the base URL of the server, this message is displayed, confirming that the server is operational. The `@app.route("/")` decorator specifies that this function should be triggered by the root URL:

```
@app.route("/")
def welcome():
    return "<p>Welcome to our Fitness Evaluation Server!</p>"
```

3. We then reuse the `busy_wait()` function from the previous chapter. This function simulates a CPU-intensive fitness evaluation, of a duration specified by the `DELAY_SECONDS` constant:

```
def busy_wait(duration):
    current_time = time.time()
    while (time.time() < current_time + duration):
        pass
```

4. Finally, we define the actual fitness evaluation function, `oneMaxFitness()`. Decorated by the `/one_max_fitness` route, it expects a value (the genetic individual) to be passed in the URL, which is then processed by the function. The function calls `busy_wait` to simulate processing, then calculates the sum of 1s in the provided string and returns this sum as a string. We use strings as input and output for this function to accommodate HTTP's text-based data transmission in web applications:

```
@app.route("/one_max_fitness/<individual_as_string>")
def oneMaxFitness(individual_as_string):
    busy_wait(DELAY_SECONDS)
    individual = [int(char) for char in individual_as_string]
    return str(sum(individual))
```

To initiate our Flask-based web application, we first need to navigate into the `server` directory, and then activate the server's virtual environment. Once activated, the application can be launched with the following command, executed from the terminal within that environment:

```
flask --app fitness_evaluator run
```

This yields the following output:

```
*  Serving Flask app 'fitness_evaluator'
*  Debug mode: off
WARNING: This is a development server. Do not use it in a production
deployment. Use a production WSGI server instead.
*  Running on http://127.0.0.1:5000
Press CTRL+C to quit
```

The warning regarding Flask's built-in server is a reminder that it's not designed for performance optimization and is meant for development purposes only. However, for our current needs, this server is perfectly appropriate to test and verify the logic of our application. To do this, we can simply use a web browser on our local machine.

Upon opening a browser and navigating to the specified URL (`http://127.0.0.1:5000`), we should see the "Welcome" message appear, indicating that our server is up and running, as illustrated here:

Welcome to our Fitness Evaluation Server!

Figure 14.2: Display of the Welcome message at the server's root URL

Next, let's test the fitness function by navigating to the following URL: `http://127.0.0.1:5000/one_max_fitness/1100110010`.

Accessing this URL internally triggers a call to the `oneMaxFitness()` function with `1100110010` as the argument. As expected, after a delay of several seconds (introduced by the `busy_wait()` function to simulate processing time), we receive a response. The browser displays the number 5, which represents the calculated sum of 1s in the input string, as illustrated in the figure here:

5

Figure 14.3: Testing the server's fitness function via a browser

Now that we've successfully set up and verified the server, let's move on to implementing the client side.

Creating the client component

To start working on the client module, we need to switch back to the original virtual environment used for our genetic algorithms' programs. This can be done by using a separate terminal or opening a new window in the IDE where this environment is activated. The various programs implementing the client module can be found at the following location:

`https://github.com/PacktPublishing/Hands-On-Genetic-Algorithms-with-Python-Second-Edition/tree/main/chapter_14`

The first program we'll examine is `01_one_max_client.py`, which functions as a straightforward (synchronous) client. This program can be found at the following location:

`https://github.com/PacktPublishing/Hands-On-Genetic-Algorithms-with-Python-Second-Edition/blob/main/chapter_14/01_one_max_client.py`

This program adapts `01_one_max_start.py` from the previous chapter—the basic OneMax problem solver. To support the delegation of fitness calculation to the server, we made the following modifications:

1. The Python `urllib` module is imported. This module provides a suite of functions and classes for working with URLs, which we will use to send HTTP requests to the server and retrieve the responses.

2. The `BASE_URL` constant is defined to point to the base URL of the server:

    ```
    BASE_URL="http://127.0.0.1:5000"
    ```

3. The `oneMaxFitness()` function is renamed `oneMaxFitness_client()`. This function converts the given individual—a list of integers (0 or 1)—into a single string. It then uses the `urlopen()` function from `urllib` to send this string to the fitness calculation endpoint on the server, by combining the base URL with the function's route and appending the string representing the individual. The function waits (synchronously) for the response and converts it back to an integer:

    ```
    def oneMaxFitness_client(individual):
        individual_as_str = ''.join(str(bit) for bit in individual)
        response = urlopen(f'{BASE_URL}/one_max_fitness/{individual_
    as_str}')
        if response.status != 200:
            print("Exception!")
        sum_digits_str = response.read().decode('utf-8')
        return int(sum_digits_str),
    ```

We can now launch this client program while the Flask server is up, and watch the server's output, showing requests coming in. It is evident that the requests are coming one at a time, with a noticeable three-second delay between them. Meanwhile, on the client side, the output mirrors what we observed in the previous chapter before introducing multiprocessing:

```
gen      nevals   max      avg
0        20       7        4.35
1        14       7        6.1
2        16       9        6.85
3        16       9        7.6
4        16       9        8.45
5        13       10       8.9
Best Individual  =   [1, 1, 1, 1, 1, 1, 1, 1, 1, 1]
Elapsed time = 285.53 seconds
```

The elapsed time is similar as well, roughly 3 seconds times the amount of fitness function invocations (95), reaffirming the synchronous nature of our current client-server interaction.

Now that the operation has been successful and the fitness function has been effectively separated and moved to the server, let's take the next step and convert the client into an asynchronous one.

Creating the asynchronous client

To support asynchronous I/O, we are going to use aiohttp, a powerful Python library for asynchronous HTTP client/server networking. This library, along with its dependencies, can be installed in the client's virtual environment using the following command:

```
pip install aiohttp
```

The modules modified for the asynchronous version of the client include not only the 02_one_max_async_client.py program but also elitism_async.py, which replaces elitism.py used in most of our programs so far. While 02_one_max_async_client.py contains the function that sends fitness calculation requests to the server, elitism_async.py manages the main genetic algorithm loop and is responsible for calling that function. The following subsections will delve into the details of these two programs.

Updating the OneMax solver

Let's start with 02_one_max_async_client.py, which can be found here:

https://github.com/PacktPublishing/Hands-On-Genetic-Algorithms-with-Python-Second-Edition/blob/main/chapter_14/02_one_max_async_client.py

The differences from the previous, synchronous program `01_one_max_client.py` are highlighted here:

1. The `oneMaxFitness_client()` function has been renamed `async_oneMaxFitness_client()`. In addition to `individual`, the new function also receives a `session` argument of the `aiohttp.ClientSession` type; this object is responsible for managing asynchronous requests while reusing connections. The function signature is preceded by the `async` keyword, marking it as a *coroutine*. This designation allows the function to pause its execution and yield control back to the event loop, enabling the concurrent sending of requests:

    ```
    async def async_oneMaxFitness_client(session, individual):
    ```

2. The HTTP GET request to the server is sent using the `session` object. When a response is received, it is available in the `response` variable:

    ```
    async with session.get(url) as response:
        . . .
    ```

3. The `main()` function, which now utilizes calls to async functions, is defined with the `async` keyword as well.

4. The call to the genetic algorithm's main loop now utilizes the `elitism_async` module instead of the original `elitism`. This module will be examined shortly. In addition, the call is preceded by the `await` keyword, required when calling an async function, to signify the function's ability to hand control back over to the event loop:

    ```
    population, _ = await elitism_async.eaSimpleWithElitism(
        population, toolbox, cxpb=P_CROSSOVER, mutpb=P_MUTATION,
        ngen=MAX_GENERATIONS, stats=stats,
        halloffame=hof, verbose=True)
    ```

5. The call to the `main()` function is made using `asyncio.run()`. This method of invocation is used to designate the main entry point for running the asynchronous program. It initiates and manages the `asyncio` event loop, thereby allowing asynchronous tasks to be scheduled and executed:

    ```
    asyncio.run(main())
    ```

Updating the genetic algorithm loop

The counterpart program, `elitism_async.py`, can be found here:

https://github.com/PacktPublishing/Hands-On-Genetic-Algorithms-with-Python-Second-Edition/blob/main/chapter_14/elitism_async.py

As previously mentioned, this program is a modified version of the familiar `elitism.py`, tailored to execute the genetic algorithm's main loop asynchronously and manage the asynchronous calls to the fitness function. The key modifications are highlighted here:

1. First, an `aiohttp.TCPConnector` object is created before the loop starts. This object is in charge of creating and managing TCP connections for sending HTTP requests. The `limit` parameter is used here to control the number of simultaneous connections to the server:

    ```
    connector = aiohttp.TCPConnector(limit=100)
    ```

2. Next, an `aiohttp.ClientSession` object is created. This session, which facilitates the sending of HTTP requests asynchronously, is used throughout the rest of the code. It's also passed to the `async_oneMaxFitness_client()` function where it is used to transmit requests to the server. The session remains active throughout the loop, enabling responses to be matched with their respective requests as they arrive:

    ```
    async with aiohttp.ClientSession(connector=connector) as
    session:
    ```

3. The original calls to the fitness evaluation function, implemented using the `map()` function to apply the `evaluate` operation to all individuals needing updated fitness values (`invalid_ind`), are now replaced with the following two lines of code. These lines create a list of `Task` objects named `evaluation_tasks`, representing scheduled calls to the asynchronous fitness function, and then wait for all of them to complete:

    ```
    evaluation_tasks = [asyncio.ensure_future(
        toolbox.evaluate(session, ind)) for ind in invalid_ind]
    fitnesses = await asyncio.gather(*evaluation_tasks)
    ```

We are now ready to use the new asynchronous client, as will be described next.

Running the asynchronous client

First, ensure that the Flask server is operational. If it's not already running, you can start it using the following command:

```
flask --app fitness_evaluator run
```

Next, let's launch the `02_one_max_async_client.py` program and observe the outputs both from the server and the client windows.

In contrast to the previous experiment, it's now evident that the requests are reaching the server several at a time and are being handled concurrently. On the client side, while the output appears similar to the previous run, the runtime is significantly improved—more than 10 times faster:

```
gen     nevals  max     avg
0       20      7       4.35
```

```
1        14      7        6.1
2        16      9        6.85
3        16      9        7.6
4        16      9        8.45
5        13      10       8.9
Best Individual =  [1, 1, 1, 1, 1, 1, 1, 1, 1, 1]
Elapsed time = 25.61 seconds
```

Now that we have learned how to utilize a client-server model for the OneMax problem, we'll learn how to use a production app server to host the models.

Using a production-grade app server

As we have mentioned before, the built-in server that comes with Flask is not optimized for performance and is intended primarily for development purposes. While it appears to perform adequately in our asynchronous experiment, when transitioning an application to a production environment, it's strongly recommended to use a production-grade **Web Server Gateway Interface** (**WSGI**) server. WSGI is a specification in Python that provides a universal interface between web servers and web applications or frameworks. Several robust options are available, including Gunicorn, uWSGI, and Apache with mod_wsgi. These servers are designed to meet the demands of a production environment, offering enhanced performance, security, stability, and scalability. As we will demonstrate next, migrating to one of these servers is a relatively easy task.

Using the Gunicorn server

The first server option we will explore is **Gunicorn**, short for **Green unicorn**. It is a widely used Python WSGI HTTP server for Unix systems, known for its simplicity and efficiency, making it a popular choice for deploying Python web applications.

Although Gunicorn is not natively supported on Windows, it can be used via the **Windows Subsystem for Linux** (**WSL**), which is supported by Windows 10 and later versions. WSL allows you to run a GNU/Linux environment directly on Windows, unmodified, without the overhead of a traditional virtual machine or dual-boot setup. Gunicorn can be installed and run within this Linux environment.

To install Gunicorn, make sure you are within the server's virtual environment, and use the following command:

```
pip install gunicorn
```

Then, launch the server with the following command:

```
gunicorn -b 127.0.0.1:5000 --workers 20 fitness_evaluator:app
```

The -b parameter, which is optional, is used here to run the server on the local URL 127.0.0.1:5000, aligning with the original Flask server configuration. By default, Gunicorn runs on port 8000.

The --workers argument specifies the number of worker processes. In the absence of this argument, Gunicorn defaults to using a single worker process.

Once the Gunicorn server is up, running the client yields the following output:

```
gen      nevals   max      avg
0        20       7        4.35
1        14       7        6.1
2        16       9        6.85
3        16       9        7.6
4        16       9        8.45
5        13       10       8.9
Best Individual =   [1, 1, 1, 1, 1, 1, 1, 1, 1, 1]
Elapsed time = 18.71 seconds
```

Recall that the best theoretical result we can achieve in this experiment is 18 seconds, as we have 6 "rounds" of fitness calculation, and the best possible result for each round is 3 seconds, the duration of a single fitness evaluation. The result we obtained here is impressively close to that theoretical limit.

In case you would like to use a server native to Windows, we will cover the Waitress server in the next subsection.

Using the Waitress server

Waitress is a production-quality pure-Python WSGI server. It's a cross-platform server compatible with various operating systems, including Unix, Windows, and macOS.

Waitress is often used as an alternative to Gunicorn, especially in environments where Gunicorn is not available or preferred, such as Windows, or when a pure-Python solution is desired.

To install Waitress, make sure you are within the server's virtual environment, and use the following command:

```
pip install waitress
```

Next, we need to make a couple of modifications to our Flask application. The modified program, fitness_evaluator_waitress.py, can be found here:

https://github.com/PacktPublishing/Hands-On-Genetic-Algorithms-with-Python-Second-Edition/blob/main/chapter_14/server/fitness_evaluator_waitress.py

The differences from the original program, `fitness_evaluator.py`, are highlighted here:

1. First, we import the `serve` function from the Waitress module:

```
from waitress import serve
```

2. Then, we use the `serve()` function to launch the server from within the program. This function allows us to specify the server's configuration through its arguments. In our case, we're setting the host, port, and the number of threads to handle requests:

```
if __name__ == "__main__":
    serve(app, host='0.0.0.0', port=5000, threads=20)
```

The server can be started by running the following program: `fitness_evaluator_waitress.py`.

Breaking out of the box

The next logical step would be to deploy the server component of your application to a separate platform. Doing so offers several key advantages, including scalability, enhanced performance, and improved reliability. While deploying your server on-premises using your own hardware is an option, leveraging **cloud computing services** often provides a more efficient and effective solution. We will cover this in more detail in the next section.

Reaching for the sky with cloud computing

Cloud computing services provide businesses and individuals with on-demand access to a wide range of applications, storage solutions, and computing power over the internet. These services eliminate the need for substantial upfront investment in physical infrastructure, allowing users to pay only for the resources they use. Cloud computing supports an extensive array of applications, ranging from data storage and web hosting to advanced analytics and artificial intelligence, revolutionizing how organizations manage and deploy IT solutions.

Additional advantages of cloud platforms include high-level security measures, data redundancy, and global reach, ensuring low latency for users worldwide. Moreover, cloud services reduce the need for upfront capital investment in hardware and minimize the burden of ongoing maintenance and upgrades. This approach enables a greater focus on application development rather than infrastructure management.

When considering the deployment of our Flask-based server component on a cloud platform, it's important to note that most major cloud service providers offer straightforward methods to deploy Flask applications. For example, a Flask application can be easily deployed to **Azure App Service**, a fully managed platform for hosting web applications provided by Microsoft's Azure cloud computing service. This platform simplifies much of the deployment and management process, making it a convenient choice for Flask deployments. Detailed instructions and guidelines for deploying a Flask application to Azure App Service can be found at this link:

`https://learn.microsoft.com/en-us/azure/app-service/quickstart-python`

Several other options are available from **Amazon Web Services** (**AWS**). You can use **Amazon EC2** for comprehensive control over virtual servers, or opt for **AWS Fargate** if you prefer a container-based compute service that doesn't require managing underlying servers. A more straightforward approach is using **AWS Elastic Beanstalk**, a user-friendly service for deploying and scaling web applications. Elastic Beanstalk automates various deployment details such as capacity provisioning, load balancing, auto-scaling, and application health monitoring. Deploying an existing Flask application to Elastic Beanstalk using the AWS **Command-Line Interface** (**CLI**) is straightforward, as described here:

```
https://docs.aws.amazon.com/elasticbeanstalk/latest/dg/create-deploy-
python-flask.html
```

In the rest of this chapter, however, our focus shifts to a fourth option—**AWS Lambda**. AWS Lambda represents a paradigm shift in application deployment and management. As a serverless computing service, it allows the execution of code without the need for provisioning or managing servers, automatically scaling in response to incoming requests. This serverless approach offers a distinct set of benefits for deploying Flask applications.

> **Important – limitations of Lambda**
>
> Before proceeding, it's essential to remember that the AWS Lambda service, while powerful and versatile, does have certain limitations and restrictions. The most significant of these is the maximum execution time limit per function invocation, currently set at 15 minutes. This means that for a genetic algorithm where a single evaluation of the fitness function is expected to exceed this duration, the method we describe next will not be suitable, and one of the aforementioned alternative approaches, such as AWS Elastic Beanstalk, should be considered.
>
> Other limitations of Lambda include constraints on memory and compute resources, deployment package size, and the number of concurrent executions, as described here:
>
> ```
> https://docs.aws.amazon.com/lambda/latest/dg/gettingstarted-
> limits.html.
> ```

Despite the aforementioned limitations, AWS Lambda remains a viable option for numerous problems tackled by genetic algorithms. In many cases, the duration required to complete a single evaluation of the fitness function falls well within the 15-minute execution time limit imposed by Lambda. Additionally, the resources provided by the Lambda service are often more than adequate for these applications. This compatibility makes AWS Lambda an attractive choice for executing genetic algorithms efficiently, which we will explore in the following sections.

AWS Lambda and API Gateway – a winning combination

AWS Lambda is a serverless computing service offered by AWS, enabling the execution of code without the need for server provisioning or management. As a prime example of **Function-as-a-Service** (**FaaS**), a cloud computing model, Lambda empowers developers to write and update code executed in response to specific events. In this model, the underlying physical hardware, server operating system maintenance, automatic scaling, and capacity provisioning are all managed by the platform, allowing developers to concentrate on individual functions in their application code. Lambda's automatic scaling adjusts the compute capacity for each trigger, ensuring high availability.

The cost-effectiveness of using AWS Lambda lies in its billing structure, which charges only for the actual compute time used, with no costs incurred when the code is not running. Moreover, Lambda's seamless integration with other AWS services makes it an invaluable tool for developing complex applications. A key integration is with **AWS API Gateway**, a fully managed service that serves as a "front door" for applications, effectively enabling API Gateway to trigger Lambda functions in response to HTTP requests. This integration facilitates the creation of serverless architectures, where Lambda functions are invoked via API calls through API Gateway.

This powerful combination allows us to deploy our existing Flask application to the AWS cloud, leveraging both the API Gateway and Lambda service. What's more, thanks to the Zappa framework, which will be described in the next section, we can deploy our Flask application without any modifications, fully utilizing the benefits of serverless architecture.

Serverless Python with Zappa

Zappa is an open source framework that simplifies the deployment of Python web applications on AWS Lambda. It's particularly well-suited for Flask (as well as Django—another Python web framework) applications. Zappa handles all of the setup and configuration required to run a web application on Lambda, turning it into a serverless application. This includes packaging the application, setting up the necessary AWS configurations, and deploying it to Lambda.

In addition, Zappa provides features such as database migrations, scheduling of function executions, and integration with various AWS services, making it a comprehensive tool for deploying Python web applications on AWS Lambda.

To install Zappa, make sure you are within the server's virtual environment, and use the following command:

```
pip install zappa
```

Before proceeding further, it's important to have a working AWS account, as described in the next subsection.

Setting up an AWS account

To be able to deploy our server to the AWS cloud, you need a valid AWS account. The **AWS Free Tier**, available to new AWS customers, allows you to explore and use AWS services for free within certain usage limits.

If you don't currently have an AWS account, you can sign up for a free account here:

```
https://aws.amazon.com/free/
```

Next, install the AWS CLI by following the instructions available here:

```
https://docs.aws.amazon.com/cli/latest/userguide/cli-chap-getting-
started.html
```

You will also need to set up your AWS credentials file, as described here:

```
https://wellarchitectedlabs.com/common/documentation/aws_credentials/
```

These will be used by Zappa behind the scenes as we continue to deploy our service.

Deploying the server module to the Lambda service

It's time to deploy our Flask application to AWS using Zappa. Navigate into the server directory, make sure the server virtual environment is active, and issue the following command:

```
zappa init
```

This will start an interactive dialog. Zappa will prompt you for various details such as the production environment name (with dev as the default), a unique S3 bucket name for storing files (it suggests a unique name for you), and the name of your application (which should be set to fitness_evaluator. app in your case). It will also ask about the global deployment option, with n as the default choice. You can typically accept all the default values provided by Zappa during this setup process.

The outcome of this initialization process is a file named zappa_settings.json. This file contains the deployment configuration for your application. If necessary, you can manually edit this file to modify the configuration or add additional options.

We are now ready to deploy our application. If you chose dev as the name for your production environment during the Zappa configuration, use the following command:

```
zappa deploy dev
```

The deployment process may take a few minutes. Once it's complete, you will see a message indicating Deployment complete! along with a URL. *This URL serves as the base URL for your newly deployed application.*

We can now manually test the deployment by pointing a browser to the new URL. The response **Welcome to our Fitness Evaluation Server!** should be displayed on the browser's screen. Next, we can also test the fitness evaluation endpoint, by adding /one_max_fitness/1100110010 to the base URL. The response, 5, should be displayed after a few seconds.

Before we continue to use our async client module with the newly deployed server, we can log in to the AWS console to review what has been deployed. This step is optional—if you're already familiar with the AWS console, feel free to skip to the following section.

Examining the deployment on AWS

To review the main components deployed by Zappa, start by logging in to the AWS console at `https://aws.amazon.com/console/`.

Once logged in, navigate to the Lambda service, where you can view the list of available Lambda functions. You should see your newly deployed function listed there, similar to the following screen capture:

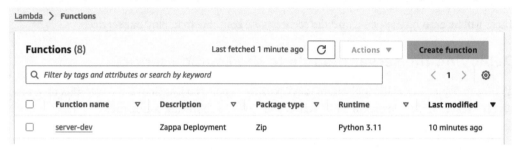

Figure 14.4: The Lambda function created by the Zappa deployment

In our case, the Lambda function created by Zappa is named `server-dev`. This name is derived from the combination of the directory name where the application resides (`server`) and the production environment name we chose (`dev`).

Clicking on the function's name will bring us to the **Function overview** screen, where we can further explore detailed information, such as the function's runtime environment, triggers, configuration settings, and monitoring metrics, as seen here:

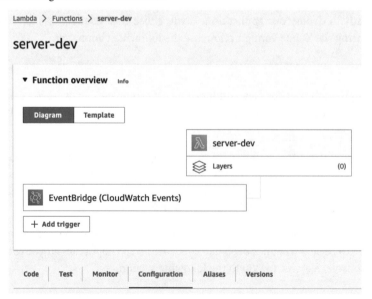

Figure 14.5: The Lambda Function overview screen

Next, let's navigate to the API Gateway service, where you can view the list of available APIs. You should see our newly deployed API listed there, with the same name as the Lambda function, as shown here:

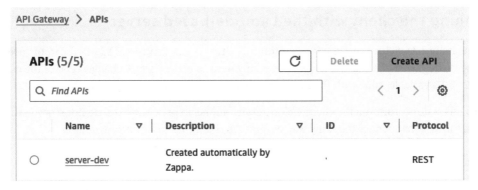

Figure 14.6: The API created by the Zappa deployment

Clicking on the API's name will bring us to the **Resources** screen; then, selecting the **ANY** link will display a diagram that illustrates how the API Gateway routes incoming requests to your Lambda function and how responses are returned to the client, as shown in the following screen capture:

Figure 14.7: The API Gateway Resources screen

When you click on the Lambda symbol on the right side, it will display the name of our Lambda function. This includes a hyperlink, which, when clicked, will take us back to the page of our Lambda function.

Running the client with the Lambda-based server

To update our asynchronous client program, `02_one_max_async_client.py`, for use with our newly deployed Lambda-based server, we only need to make a single change: replace the existing `BASE_URL` variable value with the new URL provided by the Zappa deployment.

Once this is done, running the client yields output similar to previous runs, demonstrating that the genetic algorithm operates the same way despite the change in server infrastructure:

```
gen     nevals  max     avg
0       20      7       4.35
1       14      7       6.1
2       16      9       6.85
3       16      9       7.6
4       16      9       8.45
5       13      10      8.9
Best Individual =  [1, 1, 1, 1, 1, 1, 1, 1, 1, 1]
Elapsed time = 19.54 seconds
```

Rerunning the client several times, the results indicate elapsed time values of between 19 and 21 seconds, which are reasonable considering the server is operating in a cloud environment with inherent network latency and serverless function initialization times.

Undeploying the server

Once we're finished using the server deployed via Zappa, it's a good practice to undeploy its infrastructure using the `zappa undeploy` command, initiated from within our server's virtual environment:

```
zappa undeploy dev
```

This action helps manage costs and resources efficiently by removing the deployed AWS resources that are no longer in use.

Summary

In this chapter, you learned how to restructure a genetic algorithm into a client-server model. The client uses asynchronous I/O, while the server, built with Flask, handles fitness function calculations. The server component was then successfully deployed to the cloud using Zappa, making it operational as an AWS Lambda service. This approach demonstrates the effective use of serverless computing in enhancing the performance of genetic algorithms.

In the next chapter, we'll explore how genetic algorithms can be creatively applied in the art world. Specifically, we'll learn how these algorithms can be used to reconstruct images of famous paintings using semi-transparent, overlapping shapes. This approach not only offers a unique blend of art and technology but also provides an insightful look into the versatile applications of genetic algorithms in fields beyond traditional computing.

Further reading

For more information on the topics that were covered in this chapter, please refer to the following resources:

- *Building Web Applications with Flask* by Italo Maia, June 2015

- *Expert Python Programming: Master Python by learning the best coding practices and advanced programming concepts, 4th Edition* by Michal Jaworski and Tarek Ziade, May 2021 (the *Asynchronous programming* chapter)

- *AWS Lambda Quick Start Guide: Learn how to build and deploy serverless applications on AWS* by Markus Klems, June 2018

- *Mastering AWS Lambda: Learn how to build and deploy serverless applications* by Yohan Wadia and Udita Gupta, August 2017

- Zappa framework documentation:

  `https://github.com/zappa/Zappa`

- Python `asyncio` library:

  `https://docs.python.org/3/library/asyncio.html`

Part 5:
Related Technologies

This part explores the application of genetic algorithms in image processing and introduces additional biologically inspired problem-solving techniques. The first chapter is dedicated to using genetic algorithms for image reconstruction, where an image is recreated using semi-transparent polygons, culminating in a genetic algorithm-based program that reconstructs a famous painting using these techniques. In the following chapter, the scope broadens to include genetic programming, **NeuroEvolution of Augmenting Topologies** (**NEAT**), and particle swarm optimization, each demonstrated through Python-based problem-solving programs. We conclude with an overview of several other computational paradigms in this field, further expanding our understanding of evolutionary computation methods.

This part contains the following chapters:

- *Chapter 15, Genetic Image Reconstruction*
- *Chapter 16, Other Evolutionary and Bio-Inspired Computation Techniques*

15

Evolutionary Image Reconstruction with Genetic Algorithms

In this chapter, we are going to experiment with one of the most popular ways genetic algorithms have been applied to image processing – the reconstruction of an image with a set of semi-transparent polygons. Along the way, we will gain useful experience in **image processing**, coupled with a visual insight into the evolutionary process.

We will start with an overview of image processing in Python and get acquainted with two useful libraries – *Pillow* and *OpenCV-Python*. Then, we will find out how an image can be drawn from scratch using polygons and how the difference between two images can be calculated. Next, we will develop a genetic algorithm-based program to reconstruct a segment of a famous painting using polygons and examine the results.

In this chapter, we will cover the following topics:

- Getting familiar with several image processing libraries for Python

- Understanding how to programmatically draw an image using polygons

- Finding out how to programmatically compare two given images

- Using genetic algorithms, in combination with image processing libraries, to reconstruct an image using polygons

We will start this chapter by providing an overview of the image reconstruction task.

Technical requirements

In this chapter, we will be using Python 3 with the following supporting libraries:

- `deap`
- `numpy`
- `matplotlib`
- `seaborn`
- `pillow` – introduced in this chapter
- `opencv-python (cv2)` – introduced in this chapter

> **Important note**
>
> If you use the `requirements.txt` file we provide (see *Chapter 3*), these libraries are already included in your environment.

The programs that will be used in this chapter can be found in this book's GitHub repository at `https://github.com/PacktPublishing/Hands-On-Genetic-Algorithms-with-Python-Second-Edition/tree/main/chapter_15`.

Check out the following video to see the code in action:

`https://packt.link/OEBOd`

Reconstructing images with polygons

One of the most captivating applications of genetic algorithms in image processing is reconstructing a given image using a collection of semi-transparent, overlapping shapes. This approach is not only enjoyable and educational in terms of image processing experience but also offers a compelling visual representation of the evolutionary process. Furthermore, these experiments can lead to a deeper understanding of visual arts and potentially contribute to advancements in image analysis and compression.

In these image reconstruction experiments – multiple variations of which can be found on the internet – a familiar image, often a famous painting or a fragment of it, is used as a reference. The goal is to construct a similar image by assembling a collection of overlapping shapes, typically polygons, of varying colors and transparencies.

Here, we will address this challenge by utilizing the genetic algorithms approach and the `deap` library, just like we've done for numerous types of problems throughout this book. However, since we will need to draw images and compare them to a reference image, let's get acquainted with the basics of image processing in Python.

Image processing in Python

To achieve our goal, we will need to carry out various image processing operations; for example, we will need to create an image from scratch, draw shapes onto an image, plot an image, open an image file, save an image to a file, compare two images, and possibly resize an image. In the following sections, we will explore some of the ways these operations can be performed when using Python.

Python image processing libraries

Out of the wealth of image processing libraries available for Python programmers, we chose to use two of the most prominent ones. These libraries will be briefly discussed in the following subsections.

The Pillow library

Pillow is a currently maintained fork of the original *Python Imaging Library* (PIL). It offers support for opening, manipulating, and saving image files of various formats. Since it allows us to handle image files, draw shapes, control their transparency, and manipulate pixels, we will use it as our main tool in creating the reconstructed image.

The home page of this library can be found at `https://python-pillow.org/`. A typical installation of Pillow uses the `pip` command, as follows:

```
pip install Pillow
```

The Pillow library uses the `PIL` namespace. If you have the original `PIL` library already installed, you will have to uninstall it first. More information can be found in the documentation, which is located at `https://pillow.readthedocs.io/en/stable/index.html`.

The OpenCV-Python library

OpenCV is an elaborate library that provides numerous algorithms related to computer vision and machine learning. It supports a wide variety of programming languages and is available on different platforms. *OpenCV-Python* is the Python API for this library. It combines the speed of the C++ API with the ease of use of the Python language. Here, we will mainly make use of this library to calculate the difference between two images, since it allows us to represent an image as a numeric array. We will also use its `GaussianBlur` function, which produces a blurred version of an image.

The home page of OpenCV-Python can be found here: `https://github.com/opencv/opencv-python`

The library consists of four different packages, all of which use the same namespace (`cv2`). Only one of these packages should be selected to be installed in a single environment. For our purposes, we can use the following command, which only installs the main modules:

```
pip install opencv-python
```

More information can be found in the OpenCV documentation, which is located at `https://docs.opencv.org/master/`.

Drawing images with polygons

To draw an image from scratch, we can use *Pillow*'s `Image` and `ImageDraw` classes as follows:

```
image = Image.new('RGB', (width, height))
draw = ImageDraw.Draw(image, 'RGBA')
```

`'RGB'` and `'RGBA'` are the values for the `mode` argument. The `'RGB'` value indicates three 8-bit values per pixel – one for each of the colors of red (R), green (G), and blue (B). The `'RGBA'` value adds a fourth 8-bit value (A) representing the *alpha* (opacity) level of the drawings to be added. The combination of an RGB base image and an RGBA drawing will allow us to draw polygons of varying degrees of transparency on top of a black background.

Now, we can add a polygon to the base image by using the `ImageDraw` class's `polygon` function, as shown in the following example. The following statement will draw a triangle on the image:

```
draw.polygon([(x1, y1), (x2, y2), (x3, y3)], (red, green, blue,
    alpha))
```

The following list explains the `polygon` function arguments in more detail:

- The `(x1, y1)`, `(x2, y2)`, and `(x3, y3)` tuples represent the triangle's three vertices. Each tuple contains the *x, y* coordinates of the corresponding vertex within the image.
- `red`, `green`, and `blue` are integer values in the range of [0, 255], each representing the intensity of the corresponding color of the polygon.
- `Alp~ha` is an integer value in the range of [0, 255], representing the opacity value of the polygon (a lower value means more transparency).

> **Note**
> To draw a polygon with more vertices, we would need to add more (x_i, y_i) tuples to the list.

Using the `polygon` function repetitively, we can add more and more polygons, all drawn onto the same image and possibly overlapping each other, as shown in the following figure:

Figure 15.1: A plot of overlapping polygons with varying colors and opacity values

Once we draw an image using polygons, we need to compare it to the reference image, as described in the next subsection.

Measuring the difference between images

Since we would like to construct an image that is as similar as possible to the original one, we need a way to evaluate the similarity or the difference between the two given images. The most common method to evaluate the similarity between images is the pixel-based **mean squared error** (**MSE**), which involves conducting a pixel-by-pixel comparison. This requires, of course, that both images are of the same dimensions. The MSE metric can be calculated as follows:

1. Calculate the square of the difference between each pair of matching pixels from both images. Since each pixel in the drawing is represented using three separate values – red, green, and blue – the difference for each pixel is calculated across these three dimensions.

2. Compute the sum of all these squares.

3. Divide the sum by the total number of pixels.

When both images are represented using the OpenCV (cv2) library, which essentially represents an image as a numeric array, this calculation can be performed in a straightforward manner as follows:

```
MSE = np.sum(
    (cv2Image1.astype("float") -
    cv2Image2.astype("float"))**2)/float(numPixels)
```

When the two images are identical, the MSE value will be zero. Consequently, minimizing this metric can be used as the objective of our algorithm, which will be further discussed in the next section.

Using genetic algorithms to reconstruct images

As we discussed previously, our goal in this experiment is to use a familiar image as a reference and create a second image, as similar as possible to the reference, using a collection of overlapping polygons of varying colors and transparencies. Using the genetic algorithms approach, each candidate solution is a set of such polygons, and evaluating the solution is carried out by creating an image using these polygons and comparing it to the reference image.

As usual, the first decision we need to make is how these solutions are represented. We will discuss this in the next subsection.

Solution representation and evaluation

As we mentioned previously, our solution consists of a set of polygons within the image boundaries. Each polygon has its own color and transparency. Drawing such a polygon using the Pillow library requires the following arguments:

- A list of tuple, $[(x_1,y_1), (x_2,y_2), \ldots, (x_n,y_n)]$, representing the vertices of the polygon. Each tuple contains the x, y coordinates of the corresponding vertex within the image. Therefore, the values of the x coordinates are in the range [0, image width – 1], while the values of the y coordinates are in the range [0, image height – 1].

- Three integer values in the range of [0, 255], representing the *red*, *green*, and *blue* components of the polygon's color.

- An additional integer value in the range of [0, 255], representing the *alpha* – or opacity – value of the polygon.

This means that for each polygon in our collection, we will need $\left[2 \times (polygon\_size) + 4\right]$ parameters. A *triangle*, for example, will require 10 parameters (2x3+4), while a *hexagon* will require 16 parameters (2x6+4). Consequently, a collection of triangles will be represented using a list in the following format, where every 10 parameters represent a single triangle:

$$\left[x_{11}, y_{11}, x_{12}, y_{12}, x_{13}, y_{13}, r_1, g_1, b_1, alpha_1, x_{21}, y_{21}, x_{22}, y_{22}, x_{23}, y_{23}, r_2, g_2, b_2, alpha_2, \ldots\right]$$

To simplify this representation, we will use float numbers in the range of [0, 1] for each of the parameters. Before drawing the polygons, we will expand each parameter accordingly so that it fits within its required range – image width and height for the coordinates of the vertices and [0, 255] for the colors and opacity values.

Using this representation, a collection of 50 triangles will be represented as a list of 500 float values between 0 and 1, like so:

```
[0.1499488467959301, 0.3812631075049196, 0.000439458056299,
 0.9988170920722447, 0.9975357316889601, 0.9997461395379549,
 ...
 0.9998952203500615, 0.48148512088979356, 0.083285509827404]
```

Evaluating a given solution means dividing this long list into "chunks" representing individual polygons – in the case of triangles, each chunk will have a length of 10. Then, we need to create a new, blank image and draw the various polygons from the list on top of it, one by one.

Finally, the difference between the resulting image and the original (reference) image needs to be calculated. As discussed in the previous section, this will be done using the pixel-based MSE.

This (somewhat elaborate) score evaluation procedure is implemented by a Python class, which will be described in the next subsection.

Python problem representation

To encapsulate the image reconstruction challenge, we've created a Python class called `ImageTest`. This class is contained in the `image_test.py` file, which is located at `https://github.com/PacktPublishing/Hands-On-Genetic-Algorithms-with-Python-Second-Edition/blob/main/chapter_15/image_test.py`.

The class is initialized with two parameters: the path of the file containing the reference image and the number of vertices of the polygons that are being used to construct the image. The class provides the following public methods:

- `polygonDataToImage()`: Accepts the list containing the polygon data we discussed in the previous subsection, divides this list into chunks representing individual polygons, and creates an image containing these polygons by drawing the polygons one by one onto a blank image.

- `getDifference()`: Accepts polygon data, creates an image containing these polygons, and calculates the difference between this image and the reference image using the *MSE* method.

- `blur()`: Accepts an image in PIL format, converts it to OpenCV (cv2) format, and then applies Gaussian blurring. The intensity of the blur is determined by the `BLUR_KERNEL_SIZE` constant.

- `plotImages()`: For visual comparison purposes, creates a side-by-side plot of three images:

  - The reference image (to the left)

  - The given, polygon-reconstructed image (to the right)

  - A blurred version of the reconstructed image (in the middle)

- `saveImage()`: Accepts polygon data, creates an image containing these polygons, creates a side-by-side plot of this image next to the reference image, and saves the plot in a file.

During the run of the genetic algorithm, the `saveImage()` method will be called every 100 generations in order to save a side-by-side image comparison representing a snapshot of the reconstruction process. Calling this method will be carried out by a callback function, as described in the next subsection.

Genetic algorithm implementation

To reconstruct a given image with a set of semi-transparent overlapping polygons using a genetic algorithm, we've created a Python program called `01_reconstruct_with_polygons.py`, which is located at `https://github.com/PacktPublishing/Hands-On-Genetic-Algorithms-with-Python-Second-Edition/blob/main/chapter_15/01_reconstruct_with_polygons.py`.

Since we are using a list of floats to represent a solution – the polygons' vertices, colors, and transparency values – this program is very similar to the function optimization programs we saw in *Chapter 6, Optimizing Continuous Functions*, such as the one we used for the *Eggholder function*'s optimization.

The following steps describe the main parts of this program:

1. We start by setting the problem-related constant values. `POLYGON_SIZE` determines the number of vertices for each polygon, while `NUM_OF_POLYGONS` determines the total number of polygons that will be used to create the reconstructed image:

    ```
    POLYGON_SIZE = 3
    NUM_OF_POLYGONS = 100
    ```

2. After setting the genetic algorithm constants, we continue by creating an instance of the `ImageTest` class, which will allow us to create images from polygons and compare these images to the reference image, as well as save snapshots of our progress:

    ```
    imageTest = image_test.ImageTest(MONA_LISA_PATH, POLYGON_SIZE)
    ```

3. Next, we set the upper and lower boundaries for the float values we will be searching for. As we mentioned previously, we will use float values for all our parameters and set them all to the same range, between 0.0 and 1.0, for convenience. When evaluating a solution, the values will be expanded to their actual range, and converted into integers when needed:

    ```
    BOUNDS_LOW, BOUNDS_HIGH = 0.0, 1.0
    ```

4. Since our goal is to minimize the difference between the images – the reference image and the one we are creating using polygons – we define a single objective, *minimizing* fitness strategy:

    ```
    creator.create("FitnessMin", base.Fitness, weights=(-1.0,))
    ```

5. Now, we need to create a helper function that will create random real numbers that are uniformly distributed within a given range. This function assumes that the range is the same for every dimension, as is the case in our solution:

    ```
    def randomFloat(low, up):
        return [random.uniform(l, u) for l, u in zip([low] * \
            NUM_OF_PARAMS, [up] * NUM_OF_PARAMS)]
    ```

6. Next, we use the preceding function to create an operator that randomly returns a list of floats, all in the desired range of [0, 1]:

```
toolbox.register("attrFloat", randomFloat, BOUNDS_LOW,
    BOUNDS_HIGH)
```

7. This is followed by defining an operator that fills up an individual instance using the preceding operator:

```
toolbox.register("individualCreator",
    tools.initIterate,
    creator.Individual,
    toolbox.attrFloat)
```

8. Then, we instruct the genetic algorithm to use the `getDiff()` method for fitness evaluation. This, in turn, calls the `getDifference()` method of the `ImageTest` instance. As a reminder, this method, which we described in the previous subsection, accepts an individual representing a list of polygons, creates an image containing these polygons, and calculates the difference between this image and the reference image using the *MSE* method:

```
def getDiff(individual):
    return imageTest.getDifference(individual, METRIC),
toolbox.register("evaluate", getDiff)
```

9. It's time to choose our genetic operators. For the selection operator, we will use *tournament selection* with a tournament size of 2. As we saw in *Chapter 4, Combinatorial Optimization*, this selection scheme works well in conjunction with the *elitist approach* that we plan to utilize here as well:

```
toolbox.register("select", tools.selTournament, tournsize=2)
```

10. As for the *crossover* operator (aliased with `mate`) and the *mutation* operator (`mutate`), since our solution representation is a list of floats bounded to a range, we will use the specialized continuous bounded operators provided by the DEAP framework – `cxSimulatedBinaryBounded` and `mutPolynomialBounded`, respectively – which we first saw in *Chapter 6, Optimizing Continuous Functions*:

```
toolbox.register("mate",
    tools.cxSimulatedBinaryBounded,
    low=BOUNDS_LOW,
    up=BOUNDS_HIGH,
    eta=CROWDING_FACTOR)
toolbox.register("mutate",
    tools.mutPolynomialBounded,
    low=BOUNDS_LOW,
```

```
            up=BOUNDS_HIGH,
            eta=CROWDING_FACTOR,
            indpb=1.0/NUM_OF_PARAMS)
```

11. As we have done multiple times before, we will use the *elitist approach*, where the **hall of fame** (**HOF**) members – the current best individuals – are always passed untouched to the next generation. However, this time, we're going to add a new feature to this implementation – a *callback function* that will be used to save the image every 100 generations (we will discuss this callback in more detail in the next subsection):

```
population, logbook = \
    elitism_callback.eaSimpleWithElitismAndCallback(
        population,
        toolbox,
        cxpb=P_CROSSOVER,
        mutpb=P_MUTATION,
        ngen=MAX_GENERATIONS,
        callback=saveImage,
        stats=stats,
        halloffame=hof,
        verbose=True)
```

12. At the end of the run, we print the best solution and use the `plotImages()` function to show a side-by-side visual comparison to the reference image:

```
best = hof.items[0]
print("Best Solution = ", best)
print("Best Score = ", best.fitness.values[0])
imageTest.plotImages(imageTest.polygonDataToImage(best))
```

13. In addition, we have employed the multiprocessing method of using a process pool, as demonstrated and tested in *Chapter 13, Accelerating Genetic Algorithms: The Power of Concurrency*. This approach is a straightforward way to accelerate the execution of our algorithm. It simply involves adding the following lines to encapsulate the call to `eaSimpleWithElitismAndCallback()`:

```
with multiprocessing.Pool(processes=20) as pool:
    toolbox.register("map", pool.map)
```

Before we look at the results, let's discuss the implementation of the callback function.

Adding a callback to the genetic run

To be able to save the best current image every 100 generations, we need to introduce a modification to the main genetic loop. As you may recall, toward the end of *Chapter 4, Combinatorial Optimization*, we already made one modification to deap's simple genetic algorithm main loop that allowed us to introduce the *elitist approach*. To be able to introduce that change, we created the eaSimpleWithElitism() method, which is contained in a file called elitism.py. This method was a modified version of the DEAP framework's eaSimple() method, which is contained in the algorithms.py file. We modified the original method by adding the elitism functionality, which takes the members of the HOF – the current best individuals – and passes them untouched to the next generation at every iteration of the loop. Now, for the purpose of implementing a callback, we will introduce another small modification and change the name of the method to eaSimpleWithElitismAndCallback(). We will also rename the file containing it to elitism_and_callback.py.

There are two parts to this modification, as follows:

1. The first part of the modification consists of adding an argument called callback to the main-loop method:

    ```
    def eaSimpleWithElitismAndCallback(population,
        toolbox, cxpb, mutpb, ngen, callback=None,
        stats=None, halloffame=None, verbose=__debug__):
    ```

    This new argument represents an external function that will be called after each iteration.

2. The other part is within the method. Here, the callback function is called after the new generation has been created and evaluated. The current generation number and the current best individual are passed to the callback as arguments:

    ```
    if callback:
        callback(gen, halloffame.items[0])
    ```

Being able to define a callback function that will be called after each generation may prove useful in various situations. To take advantage of it here, we'll define the saveImage() function back in our 01_reconstruct_with_polygons.py program. We will use it to save a side-by-side image of the current best image and the reference image, every 100 generations, as follows:

1. We use the *modulus* (%) operator to activate the method only once every 100 generations:

    ```
    if gen % 100 == 0:
    ```

2. If this is one of these generations, we create a folder for the images if one does not exist. The folder's name references the polygon size and the number of polygons – for example, run-3-100 or run-6-50, under the images/results/ path:

    ```
    RESULTS_PATH = os.path.join(BASE_PATH, "results",
        f"run-{POLYGON_SIZE}-{NUM_OF_POLYGONS}")
    ...
    if not os.path.exists(RESULTS_PATH):
        os.makedirs(RESULTS_PATH)
    ```

3. Next, we save the image of the best current individual in that folder. The name of the image contains the number of generations that have been passed – for example, `after-300-generations.png`:

```
imageTest.imageTest.saveImage(polygonData,
    os.path.join(RESULTS_PATH, f"after-{gen}-gen.png"),
    f"After {gen} Generations")
```

We are finally ready to run this algorithm with reference images and check out the results.

Image reconstruction results

To test our program, we will use a section of the famous Mona Lisa portrait by *Leonardo da Vinci*, considered the most well-known painting in the world, as seen here:

Figure 15.2: Head crop of the Mona Lisa painting

Source: `https://commons.wikimedia.org/wiki/File:Mona_Lisa_headcrop.jpg`

Artist: Leonardo da Vinci. Licensed under Creative Commons CC0 1.0: `https://creativecommons.org/publicdomain/zero/1.0/`

Before proceeding with the program, it's important to note that the extensive polygon data and complex image processing operations involved make the running time for our genetic image reconstruction experiments significantly longer than other programs tested earlier in this book. These experiments could take several hours each to complete.

We will begin our image reconstruction using 100 triangles as the polygons:

```
POLYGON_SIZE = 3
NUM_OF_POLYGONS = 100
```

The algorithm will run for 5,000 generations with a population size of 200. As discussed earlier, a side-by-side image comparison is saved every 100 generations. At the end of the run, we can review these saved images to observe the evolution of the reconstructed image.

The following figure showcases various milestones from the resulting side-by-side saved images. As mentioned before, the middle image in each row presents a blurred version of the reconstructed image. This blurring aims to soften the sharp corners and straight lines that are typical of polygon-based reconstructions, creating an effect akin to squinting when viewing the image:

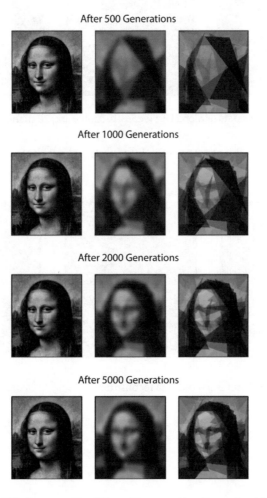

Figure 15.3: Milestone results of Mona Lisa reconstruction using 100 triangles

The end result bears a close resemblance to the original image and can be readily recognized as the Mona Lisa.

Reducing the triangle count

It is reasonable to assume that the results would be even better when increasing the number of triangles. But what if we wanted to *minimize* this number? If we reduce the number of triangles to 20, we might still be able to tell that this is the Mona Lisa, as the following results show:

Figure 15.4: Results of Mona Lisa reconstruction using 20 triangles and MSE

However, when the triangle count is further reduced to 15, the results are no longer recognizable, as seen here:

Figure 15.5: Results of Mona Lisa reconstruction using 15 triangles and MSE

A possible way to improve these results is described in the next subsection.

Blurring the fitness

Since the reconstruction becomes significantly cruder when the triangle count is low, perhaps we can improve this result by basing the fitness on the similarity between the original image and the *blurred version* of the reconstructed image, which is less crude. To try this out, we've created a slightly modified version of the original Python program, called 02_reconstruct_with_polygons_blur.py, which is located at https://github.com/PacktPublishing/Hands-On-Genetic-Algorithms-with-Python-Second-Edition/blob/main/chapter_15/02_reconstruct_with_polygons_blur.py.

The modifications are highlighted as follows:

1. The image comparison results of this program are saved into a separate directory called blur.

2. The fitness function calculation now includes an optional argument, blur=True, when calling the getDifference() function. Consequently, this function will call getMseBlur() instead of the original getMse(). The getMseBlur() function blurs the given image before calculating the MSE:

```
def getMseBlur(self, image):
    return np.sum(
        (self.blur(image).astype("float") -
         self.refImageCv2.astype("float")) ** 2
    ) / float(self.numPixels)
```

The results of running this program for 20 triangles are shown in the following figure:

Figure 15.6: Results of Mona Lisa reconstruction using 20 triangles and MSE with blur

Meanwhile, the results for 15 triangles are shown here:

Figure 15.7: Results of Mona Lisa reconstruction using 15 triangles and MSE with blur

The resulting images appear more recognizable, which makes this method a potentially viable way to achieve a lower polygon count.

Other experiments

There are many variations that you can explore. One straightforward variation is increasing the number of vertices in the polygons. We anticipate more accurate results from this approach, as the shapes become more versatile. However, it's important to note that the size of the individual polygons grows, which typically necessitates a larger population and/or more generations to achieve reasonable results.

Another interesting variation is to apply the "blur" fitness, previously used to minimize the number of polygons, to a large polygon count. This approach might lead to a somewhat "erratic" reconstruction, which is then smoothed by the blur function. The following result illustrates this, using 100 hexagons with 400 individuals and 5,000 generations, employing the "blur" MSE-based fitness:

Figure 15.8: Results of Mona Lisa reconstruction using 100 hexagons and MSE with blur

There are many other possibilities and combinations to experiment with, such as the following:

- Increasing the number of polygons

- Changing the population size and the number of generations

- Using non-polygonal shapes (such as circles or ellipses) or regular shapes (such as squares or equilateral triangles)

- Using different types of reference images (including paintings, drawings, photos, and logos)

- Opting for grayscale images instead of colored ones

Have fun creating and experimenting with your own variations!

Summary

In this chapter, you were introduced to the popular concept of reconstructing existing images with overlapping, semi-transparent polygons. You explored various image processing libraries in Python, learning how to programmatically create images from scratch using polygons and calculate the difference between two images. Subsequently, we developed a genetic algorithm-based program to reconstruct a segment of a famous painting using polygons and explored several variations in the process. We also discussed numerous possibilities for further experimentation.

In the next chapter, we will describe and demonstrate several problem-solving techniques related to genetic algorithms, as well as other biologically inspired computational algorithms.

Further reading

For more information about the topics that were covered in this chapter, please refer to the following resources:

- *Hands-On Image Processing with Python*, Sandipan Dey, November 30, 2018

- *Grow Your Own Picture*: `https://chriscummins.cc/s/genetics`

- *Genetic Programming: Evolution of Mona Lisa*: `https://rogerjohansson.blog/2008/12/07/genetic-programming-evolution-of-mona-lisa/`

16

Other Evolutionary and Bio-Inspired Computation Techniques

In this chapter, you will broaden your horizons and discover several new problem-solving and optimization techniques related to genetic algorithms. Three different techniques of this extended family – **genetic programming**, **NeuroEvolution of Augmenting Topologies (NEAT)**, and **particle swarm optimization** – will be then demonstrated through the implementation of problem-solving Python programs. Finally, we will provide a brief overview of several other related computation paradigms.

This chapter will cover the following topics:

- The evolutionary computation family of algorithms
- Understanding the concepts of **genetic programming** and how they differ from genetic algorithms
- Using genetic programming to solve the **even parity check** problem
- Understanding the concepts of **NEAT** and how they differ from genetic algorithms
- Using NEAT to solve the even parity check problem
- Understanding the concepts of particle swarm optimization
- Using particle swarm optimization to optimize **Himmelblau's function**
- Understanding the principles behind several other evolutionary and **biologically inspired techniques**

We will start this chapter by unveiling the extended family of **evolutionary computation** and discussing the main characteristics shared by its members.

Technical requirements

In this chapter, we will be using Python 3 alongside the following supporting libraries:

- `deap`
- `numpy`
- `networkx`
- `neatpy` – introduced in this chapter
- `pygame`

> **Important note**
>
> If you're using the `requirements.txt` file we've provided (see *Chapter 3*), these libraries will already be in your environment.

The programs that will be used in this chapter can be found in this book's GitHub repository at `https://github.com/PacktPublishing/Hands-On-Genetic-Algorithms-with-Python-Second-Edition/tree/main/chapter_16`.

Check out the following video to see the Code in Action:

`https://packt.link/OEBOd`.

Evolutionary computation and bio-inspired computing

Throughout this book, we have covered the problem-solving technique known as *genetic algorithms* and applied it to numerous types of problems, including **combinatorial optimization**, **constraint satisfaction,** and **continuous function optimization**, as well as to **machine learning** and **artificial intelligence**. However, as we mentioned in *Chapter 1, An Introduction to Genetic Algorithms*, genetic algorithms are just one branch within a larger family of algorithms called **evolutionary computation**. This family consists of various related problem-solving and optimization techniques, all of which draw inspiration from Charles Darwin's theory of natural evolution.

The main characteristics that are shared by these techniques are as follows:

- The starting point is an initial set (**population**) of candidate solutions.
- The candidate solutions (**individuals**) are updated iteratively to create new generations.
- Creating a new generation involves removing less successful individuals (**selection**), as well as introducing small random changes (**mutations**) to some individuals. Other operators, such as interaction with other individuals (**crossover**), may also be applied.
- As a result, as generations go by, the **fitness** of the population increases; in other words, the candidate solutions become better at solving the problem at hand.

More broadly, since evolutionary computation techniques are based on various biological systems or behaviors, they generally overlap with the algorithm family known as **bio-inspired computing**.

In the following sections, we will cover some of the most frequently used members of evolutionary computation and bio-inspired computing – some will be covered in greater detail, while the others will only be mentioned briefly. We will start by providing a detailed account of a fascinating technique that allows us to evolve actual computer programs: **genetic programming**.

Genetic programming

Genetic programming is a special form of genetic algorithm – that is, the technique we have been applying throughout this book. In this special case, the candidate solutions – or individuals – that we are evolving to find the best one for our purpose are computer programs, hence the name. In other words, when we apply genetic programming, we evolve *computer programs* to find a program that will excel at performing a particular task.

As you may recall, genetic algorithms use a representation of the candidate solutions, often referred to as a *chromosome*. This representation is subject to genetic operators, namely *selection*, *crossover*, and *mutation*. Applying these operators to the current generation results in a new generation of solutions that is expected to produce better results than its predecessor. In most of the problems we have looked at so far, this representation was a list (or an array) of values of a certain type, such as integers, Booleans, or floats. To represent a *program*, however, we typically use a *tree structure*, as shown in the following diagram:

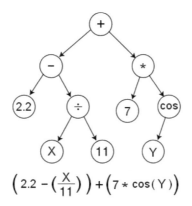

$$\left(2.2 - \left(\frac{X}{11} \right) \right) + \left(7 * \cos(Y) \right)$$

Figure 16.1: Tree structure representation of a simple program

Source: https://commons.wikimedia.org/wiki/File:Genetic_Program_Tree.png

Image by Baxelrod.

The tree structure depicted in the preceding diagram represents the calculation shown underneath the tree. This calculation is equivalent to a short program (or a function) that accepts two arguments, *X* and *Y*, and returns a certain output based on their values. To create and evolve such tree structures, we need to define two different sets:

- **Terminals**, or the *leaves* of the tree. These are arguments and the constant values that can be used in the tree. In our example, X and Y are arguments, while 2.2, 11, and 7 are constants. Constants can also be generated randomly, within a certain range, when a tree is created.

- **Primitives**, or the *internal nodes* of the tree. These are functions (or operators) that accept one or more arguments and generate a single output value. In our example, +, -, *, and ÷ are primitives that accept two arguments, while cos is a primitive that accepts a single argument.

In *Chapter 2, Understanding the Key Components of Genetic Algorithms*, we demonstrated how the genetic operator of *single-point crossover* operates on binary-valued lists. The crossover operation created two offspring from two parents by cutting out a part of each parent and swapping the detached parts between the parents. Similarly, a crossover operator for the tree representation may detach a *subtree* (a branch or a group of branches) from each parent and swap the detached branches between the parents to create offspring trees, as demonstrated in the following diagram:

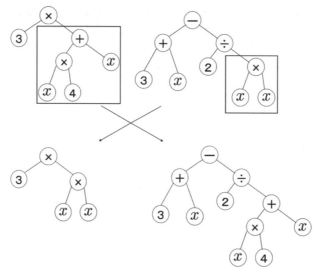

Figure 16.2: Crossover operation between two tree structures representing programs

Source: https://commons.wikimedia.org/wiki/File:GP_crossover.png

Image by U-ichi

In this example, the two parents on the top row have subtrees that have been swapped between them to create the two offspring in the second row. The swapped subtrees are marked by the rectangles surrounding them.

Along the same lines, the *mutation* operator, which intends to introduce random changes to a single individual, can be implemented by picking a subtree within the candidate solution and replacing it with a randomly generated one.

The **deap** library, which we have been using throughout this book, provides inherent support for genetic programming. In the next section, we will implement a simple genetic programming example using this library.

Genetic programming example – even parity check

For our example, we will use genetic programming to create a program that implements an even parity check. In this task, the possible values of the inputs are 0 or 1. The output value should be 1 if the number of the inputs with the value 1 is odd, thereby producing a total even number of 1 values; otherwise, the output value should be 0. The following table lists the various possible combinations of input values for the case of three inputs, along with the matching even parity output values:

in_0	in_1	in_2	Even Parity
0	0	0	0
0	0	1	1
0	1	0	1
0	1	1	0
1	0	0	1
1	0	1	0
1	1	0	0
1	1	1	1

Table 16.1: Truth table of even parity as a function of three inputs

This kind of table is often referred to as the *truth table* of the operation at hand. As evident from this truth table, one reason that the parity check is often used as a benchmark is that any single change in the input values will result in a change to the output value.

The parity check can also be represented using logic gates, such as AND, OR, NOT, and exclusive OR (XOR). While the NOT gate accepts a single input and inverts it, each of the three other gate types accepts two inputs. For the respective output to be 1, the AND gate requires both inputs to be 1, the

OR gate requires at least one of them to be 1, and the XOR gate requires that exactly one of them is 1, as shown in the following table:

in_0	in_1	AND	OR	XOR
0	0	0	0	0
0	1	0	1	1
1	0	0	1	1
1	1	1	1	0

Table 16.2: Truth tables of AND, OR and XOR operations of two inputs

There are many possible ways to implement the three-input parity check using logic gates. The simplest way to do this is by using two XOR gates, as shown in the following diagram:

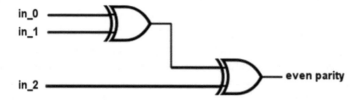

Figure 16.3: A three-input even parity check implemented using two XOR gates

In the next subsection, we will use genetic programming to create a small program that implements an even parity check using the AND, OR, NOT, and XOR logic operations.

Genetic programming implementation

To evolve a program that implements the even parity check logic, we've created a genetic programming-based Python program called 01_gp_even_parity.py at https://github.com/PacktPublishing/Hands-On-Genetic-Algorithms-with-Python-Second-Edition/blob/main/chapter_16/01_gp_even_parity.py.

Since genetic programming is a special case of genetic algorithms, much of this program will look familiar to you if you have gone over the programs we presented in earlier chapters of this book.

The following steps describe the main parts of this program:

1. We start by setting the problem-related constant values. Here, NUM_INPUTS determines the number of inputs for the even parity checker. We will use a value of 3 for simplicity; however, larger values can be set as well. The NUM_COMBINATIONS constant represents the number of possible combinations of values for the inputs, which is analogous to the number of rows in the truth table we saw earlier:

    ```
    NUM_INPUTS = 3
    NUM_COMBINATIONS = 2 ** NUM_INPUTS
    ```

2. This is followed by the familiar genetic algorithm constants we have seen numerous times before:

```
POPULATION_SIZE = 60
P_CROSSOVER = 0.9
P_MUTATION = 0.5
MAX_GENERATIONS = 20
HALL_OF_FAME_SIZE = 10
```

3. However, genetic programming requires several additional constants that refer to the tree representation of the candidate solutions. These are defined in the following code. We will see how they are used as we examine the rest of this program:

```
MIN_TREE_HEIGHT = 3
MAX_TREE_HEIGHT = 5
MUT_MIN_TREE_HEIGHT = 0
MUT_MAX_TREE_HEIGHT = 2
LIMIT_TREE_HEIGHT = 17
```

4. Next, we calculate the *truth table* of the even parity check so that we can use it as a reference when we need to check the accuracy of a given candidate solution. The `parityIn` matrix represents the input columns of the truth table, while the `parityOut` vector represents the output column. The Python `itertools.product()` function is an elegant replacement for nested `for` loops that would be otherwise required to iterate over all the combinations of input values:

```
parityIn = list(itertools.product([0, 1], repeat=NUM_INPUTS))
parityOut = [sum(row) % 2 for row in parityIn]
```

5. Now, it is time to create the set of *primitives* – that is, the operators that will be used in our evolved programs. The first declaration creates a set using the following three arguments:

    -   The name of the program to be generated using the primitives from the set (here, we called it `main`)

    -   The number of inputs to the program

    -   The prefix to be used when naming the inputs (optional)

    These three arguments are used to create the following primitive set:

```
primitiveSet = gp.PrimitiveSet("main", NUM_INPUTS, "in_")
```

6.   Now, we fill the primitive set with the various functions (or operators) that will be used as the building blocks of the program. For each operator, we will use a reference to the function that implements it and the number of arguments it expects. Although we could define our own functions for this purpose, in this case, we're making use of the existing Python `operator` module, which contains numerous useful functions, including the logical operators we need:

```
primitiveSet.addPrimitive(operator.and_, 2)
primitiveSet.addPrimitive(operator.or_, 2)
primitiveSet.addPrimitive(operator.xor, 2)
primitiveSet.addPrimitive(operator.not_, 1)
```

7.   The following definitions set the *terminal* values to be used. As we mentioned earlier, these are constants that can be used as input values for the tree. In our case, it makes sense to use 0 and 1 as values:

```
primitiveSet.addTerminal(1)
primitiveSet.addTerminal(0)
```

8.   Since our goal is to create a program that implements the truth table of the *even parity check*, we will attempt to minimize the difference between the program's output and the known output values. For this purpose, we will define a single objective – that is, minimizing the fitness strategy:

```
creator.create("FitnessMin", base.Fitness, weights=(-1.0,))
```

9.   Now, we will create the `Individual` class, based on the `PrimitiveTree` class provided by the `deap` library:

```
creator.create("Individual", gp.PrimitiveTree, \
    fitness=creator.FitnessMin)
```

10. To help us construct an individual in the population, we will create a helper function that will generate random trees using the primitive set we defined earlier. Here, we're making use of the `genFull()` function offered by `deap` and providing it with the primitive set, as well as with the values for defining the minimum and maximum height of the generated trees:

```
toolbox.register("expr",
                 gp.genFull,
                 pset=primitiveSet,
                 min_=MIN_TREE_HEIGHT,
                 max_=MAX_TREE_HEIGHT)
```

11. This is followed by defining two operators, the first of which creates an individual instance using the preceding helper operator. The other generates a list of such individuals:

```
toolbox.register("individualCreator",
                 tools.initIterate,
```

```
                     creator.Individual,
                     toolbox.expr)
toolbox.register("populationCreator",
                     tools.initRepeat,
                     list,
                     toolbox.individualCreator)
```

12. Next, we create an operator to *compile* a given primitive tree into Python code using the `compile()` function offered by `deap`. Consequently, we'll use this compile operator in a function we'll create, called `parityError()`. For a given individual – a tree representing an expression – this function counts the number of rows in the truth table for which the result of the calculation differs from the expected one:

```
toolbox.register("compile", gp.compile, pset=primitiveSet)
def parityError(individual):
    func = toolbox.compile(expr=individual)
    return sum(func(*pIn) != pOut for pIn,
        pOut in zip(parityIn, parityOut))
```

13. Then, we must instruct the genetic programming algorithm to use the `getCost()` function for fitness evaluation. This function returns the parity error we just saw in tuple form that's required by the underlying evolutionary algorithm:

```
def getCost(individual):
    return parityError(individual),
toolbox.register("evaluate", getCost)
```

14. It's time to choose our genetic operators, starting with the *selection* operator (aliased with `select`). For genetic programming, this operator is typically the same *tournament selection* we have been using throughout this book. Here, we're using it with a tournament size of 2:

```
toolbox.register("select", tools.selTournament, tournsize=2)
```

15. As for the *crossover* operator (aliased with `mate`), we will use the specialized genetic programming `cxOnePoint()` operator that's provided by `deap`. Since the evolving programs are represented by trees, this operator takes two parent trees and exchanges sections of them to create two valid offspring trees:

```
toolbox.register("mate", gp.cxOnePoint)
```

16. Next is the *mutation* operator, which introduces random changes to an existing tree. The mutation is defined in two stages. First, we specify a helper operator that utilizes the specialized genetic programming `genGrow()` function, provided by `deap`. This operator creates a subtree within the limits defined by the two constants. Then, we define the mutation operator itself (aliased

with `mutate`). This operator utilizes DEAP's `mutUniform()` function, which randomly replaces a subtree in a given tree with a random one that was generated using the helper operator:

```
toolbox.register("expr_mut",
                 gp.genGrow,
                 min_=MUT_MIN_TREE_HEIGHT,
                 max_=MUT_MAX_TREE_HEIGHT)
toolbox.register("mutate",
                 gp.mutUniform,
                 expr=toolbox.expr_mut,
                 pset=primitiveSet)
```

17. To prevent individuals in the population from growing into overly large trees, potentially containing an excessive number of primitives, we need to introduce *bloat control* measures. We can do this using DEAP's `staticLimit()` function, which imposes a tree height restriction on the results of the *crossover* and *mutation* operations:

```
toolbox.decorate("mate",
    gp.staticLimit(
        key=operator.attrgetter("height"),
        max_value=LIMIT_TREE_HEIGHT))
toolbox.decorate("mutate",
    gp.staticLimit(
        key=operator.attrgetter("height"),
        max_value=LIMIT_TREE_HEIGHT))
```

18. The program's main loop is very similar to the ones we saw in earlier chapters. After creating the initial population, defining the statistics measurements, and creating the HOF object, we call the evolutionary algorithm. Like we've done multiple times before, we must apply the *elitist approach*, where the HOF members – the current best individuals – are always passed untouched to the next generation:

```
population, logbook = elitism.eaSimpleWithElitism(
    population,
    toolbox,
    cxpb=P_CROSSOVER,
    mutpb=P_MUTATION,
    ngen=MAX_GENERATIONS,
    stats=stats,
    halloffame=hof,
    verbose=True)
```

19. At the end of the run, we print the best solution, as well as the height of the tree that's being used to represent it, and its length – that is, the total number of operators contained in the tree:

```
best = hof.items[0]
print("-- Best Individual = ", best)
print(f"-- length={len(best)}, height={best.height}")
print("-- Best Fitness = ", best.fitness.values[0])
```

20. The last thing we need to do is plot a graphic illustration of the tree representing the best solution. To that end, we must utilize the graph and networks library known as *NetworkX* (nx), which we introduced in *Chapter 5, Constraint Satisfaction*. We start by calling the graph() function provided by deap, which breaks down the individual tree into the nodes, edges, and labels that are required for the graph, and then create the graph using the appropriate networkx functions:

```
nodes, edges, labels = gp.graph(best)
g = nx.Graph()
g.add_nodes_from(nodes)
g.add_edges_from(edges)
pos = nx.spring_layout(g)
```

21. Then, we draw the nodes, edges, and labels. Since the layout of this graph is not a classic hierarchical tree, we must distinguish the top node by coloring it red and enlarging it:

```
nx.draw_networkx_nodes(g, pos, node_color='cyan')
nx.draw_networkx_nodes(g, pos, nodelist=[0],
    node_color='red', node_size=400)
nx.draw_networkx_edges(g, pos)
nx.draw_networkx_labels(g, pos, **{"labels": labels,
    "font_size": 8})
```

When running this program, we get the following output:

```
gen nevals min avg
0 60 2 3.91667
1 50 1 3.75
2 47 1 3.45
. . .
5 47 0 3.15
. . .
20 48 0 1.68333
-- Best Individual = xor(and_(not_(and_(in_1, in_2)), not_(and_
(1, in_2))), xor(or_(xor(in_1, in_0), and_(0, 0)), 1))
-- length=19, height=4
-- Best Fitness = 0.0
```

Since this is a simple problem, the fitness has quickly reached the minimum value of 0, which means we were able to find a solution that correctly reproduces the *even parity check* truth table. However, the resulting expression, which consists of 19 elements and four levels in the hierarchy, seems overly complex. This is illustrated by the following plot that was produced by the program:

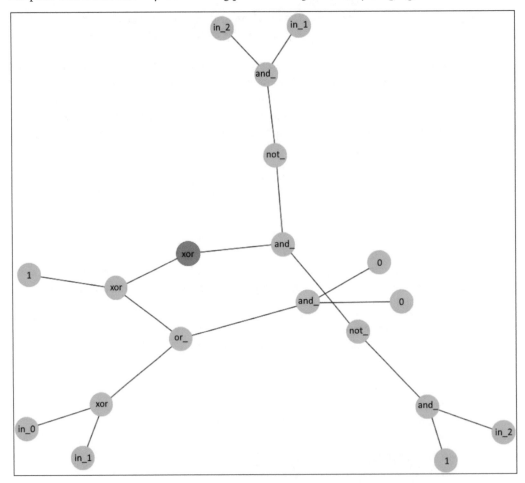

Figure 16.4: A plot representing the parity check solution that was found by the initial program

As we mentioned previously, the red node in the graph represents the top of the program's tree, which maps to the first XOR operation in the expression.

The reason for this relatively complex graph is that there is no advantage to using simpler expressions. So long as they fall within the imposed limitation of tree height, the expressions that are evaluated incur no penalty for complexity. In the next subsection, we will attempt to change this situation by making a small modification to the program in the hope of achieving the same outcome – the implementation of the even parity check – but with a simpler solution.

Simplifying the solution

In the implementation we have just seen, there were measures in place to restrict the size of the trees that represent the candidate solutions. However, the best solution we found seems overly complex. One way to pressure the algorithm into producing simpler results is to impose a small cost penalty for complexity. This penalty should be small enough that it refrains from favoring simpler solutions that fail to solve the problem. Rather, it should serve as a tie-breaker between two good solutions, so the simpler of the two will be preferred. This approach has been implemented in the `02_gp_even_parity_reduced.py` Python program, which is located at `https://github.com/PacktPublishing/Hands-On-Genetic-Algorithms-with-Python-Second-Edition/blob/main/chapter_16/02_gp_even_parity_reduced.py`.

This program is nearly identical to the previous one, except for a couple of small changes:

1. The main change was introduced to the *cost function*, which the algorithm seeks to minimize. To the original calculated error, a small penalty measure was added that depends on the height of the tree:

```python
def getCost(individual):
    return parityError(individual) + \
                individual.height / 100,
```

2. The only other change was at the end of the run, after printing the best solution that was found. Here, in addition to printing the fitness value, we print the actual *parity error* that was obtained, without the penalty that's present in the fitness:

```python
print("-- Best Parity Error = ", parityError(best))
```

By running this modified version, we get the following output:

```
gen nevals min avg
0 60 2.03 3.9565
1 50 2.03 3.7885
. . .
5 47 0.04 3.45233
. . .
10 48 0.03 3.0145
. . .
15 49 0.02 2.57983
. . .
20 45 0.02 2.88533
-- Best Individual = xor(xor(in_0, in_1), in_2)
-- length=5, height=2
-- Best Fitness = 0.02
-- Best Parity Error = 0
```

From the preceding output, we can tell that, after five generations, the algorithm was able to find a solution that correctly reproduces the even parity check truth table since the fitness value at that point was nearly 0. However, as the algorithm kept running, the tree height was reduced from four (a penalty of 0.04) to two (a penalty of 0.02). As a result, the best solution is very simple and consists of only five elements – the three inputs and two XOR operators. The solution we found represents the simplest known solution that we saw earlier, which consists of two XOR gates. This is illustrated by the following plot, which was produced by the program:

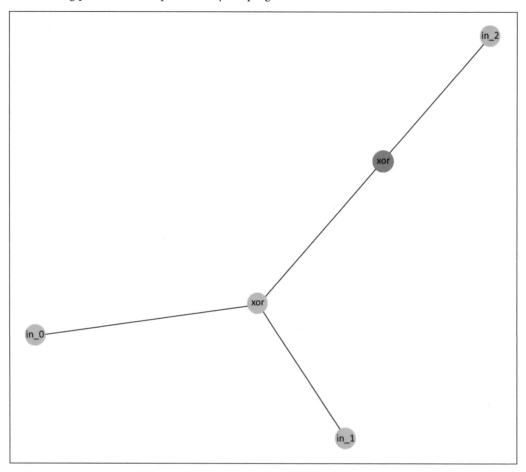

Figure 16.5: A plot representing the parity check solution that was found by the modified program

While genetic programming is considered a subset of genetic algorithms, the next section describes a more specialized form of evolutionary computation – one that is dedicated to creating neural network architectures.

NEAT

In *Chapter 9, Architecture Optimization of Deep Learning Networks*, we demonstrated how a simple genetic algorithm can be used to find the best architecture of a feed-forward neural network (also known as **multilayer perceptron** or **MLP**) for a particular task. To do that, we limited ourselves to three hidden layers and coded each network using a fixed-size chromosome that had placeholders for each of the layers, where a 0 or a negative value meant that the layer did not exist.

Taking this idea further, **NEAT** is an evolutionary technique dedicated to creating neural networks more flexibly and incrementally and was created in 2002 by *Kenneth Stanley* and *Risto Miikkulainen*.

NEAT starts with small, simple neural networks and allows them to evolve by adding and modifying neurons and connections over generations. Rather than using a fixed-size chromosome, NEAT represents solutions as *directed graphs* that directly map into artificial neural networks, where nodes represent neurons, and connections between nodes represent synapses. This allows NEAT to evolve not only the weights of the connections but also the network's structure itself, including adding and removing neurons and connections.

NEAT's *crossover* operator is designed specifically for neural networks. It aligns and combines matching neurons and connections from parent networks while maintaining unique 'innovation' identifiers. To enable this kind of matching, the history of genes is tracked by the use of a **global innovation number**, which increases as new genes are added.

In addition, NEAT employs a **speciation** mechanism that groups individuals (neural networks) into species based on their structural similarity. This grouping encourages competition *within* species rather than *between* species. This mechanism helps ensure that innovations have a chance to thrive within their respective niches before being subjected to intense competition.

NEAT (along with other related neuroevolutionary techniques) has been applied in many areas, including financial forecasting, drug discovery, evolving art, electronic circuit design, and robotics; however, it is most commonly found in *reinforcement learning* applications, such as game playing.

NEAT example – even parity check

We will illustrate using the NEAT technique by solving the same three-input *even parity check* problem we used in the previous section to demonstrate genetic programming. Here, we'll employ NEAT to create a feed-forward neural network implementation of the same parity check function.

In regards to neural networks, the even parity check, also known as **the XOR problem**, is known to be impossible for a single perceptron to implement as it forms a pattern that cannot be separated by a single line or a simple linear function. To capture this non-linearity, the minimal required network consists of, in addition to the input and output layers, a hidden layer of two neurons. In the next subsection, we will set out to see if NEAT can find this minimal solution.

NEAT implementation

To evolve a neural network that implements the even parity check logic using the NEAT technique, we've created a Python program called `03_neat_even_parity.py` at `https://github.com/PacktPublishing/Hands-On-Genetic-Algorithms-with-Python-Second-Edition/blob/main/chapter_16/03_neat_even_parity.py`.

Python NEAT library

There are several capable Python libraries available that implement the NEAT technique, most notably the **NEAT-Python** library. However, for our example, we will be using the lightweight **neatpy** library, owing to its conciseness and ease of use. This library can be installed (if not already present) using the following command:

```
pip install neatpy
```

In addition, the **PyGame** library is required for visualizing the progress of the solution. If it's not been installed yet, it can be added using the following command:

```
pip install pygame
```

Program

The following steps describe the main parts of this program:

1. Similar to the genetic programming example, we'll start by setting the problem-related constant values. `NUM_INPUTS` determines the number of inputs for the even parity checker.

2. Since we would like to save an image with the best solution's network structure at the end of the program, let's make sure a folder for it has been created:

    ```
    IMAGE_PATH = os.path.join(
        os.path.dirname(os.path.realpath(__file__)),
        "images")
    if not os.path.exists(IMAGE_PATH):
        os.makedirs(IMAGE_PATH)
    ```

3. Now, we must set up the graphical display for the real-time "animation" of the algorithm's progress by using the functionality of the *PyGame* library:

    ```
    pg.init()
    screen = pg.display.set_mode((400, 400))
    screen.fill(colors['lightblue'])
    ```

4. Next, we must set several options for the NEAT algorithm that will be used:

    - The number of inputs for our network (which would be identical to NUM_INPUTS).

    - The number of outputs (1, in our case).

    - Population size (150, in our example).

    - A fitness threshold. If the best solution surpasses this value, the algorithm considers the problem as solved and stops. As the best fitness possible is equal to the number of rows in the truth table (indicating we got the correct results for all rows), we must set the threshold to a value just under that:

    ```
    Options.set_options(NUM_INPUTS, 1, 150, 2**NUM_INPUTS - 0.1)
    ```

5. Next, we must calculate parityIn and parityOut while implementing the inputs and outputs of the desired parity check, similar to what we did in the genetic programming example:

    ```
    parityIn = list(itertools.product([0, 1], repeat=NUM_INPUTS))
    parityOut = [sum(row) % 2 for row in parityIn]
    ```

6. Now, it's time to define parityScore(), the function that evaluates a given neural network (represented by the nn parameter). Since the score needs to be positive, we'll start from the maximum score, and then subtract the square of the difference between each expected network output and the actual (float) value produced by the network:

    ```
    score = 2**NUM_INPUTS
    for pIn, pOut in zip(parityIn, parityOut):
        output = nn.predict(pIn)[0]
        score-= (output - pOut) ** 2
    ```

    In addition, the score includes a small penalty term for each node in the network, giving smaller architectures an advantage:

    ```
    score -= len(nn.nodes) * 0.01
    ```

7. Coming up next is another utility function, draw_current(). It draws the architecture (nodes and connections) of the current best solution by calling the neatpy library's draw_brain_pygame(); in addition, it illustrates the *speciation* mechanism by drawing the current status of species using the draw_species_bar_pygame() function.

8. After creating the initial population, we get to the main loop of the NEAT algorithm. Thanks to the simplicity of the neatpy library, this loop is very concise. It starts by scoring the current population, as is the usual case for evolutionary algorithms:

    ```
    for nn in p.pool:
        nn.fitness = parityScore(nn)
    ```

9. The main loop continues by calling the library's `epoch()` function, which performs a single NEAT evolutionary step, resulting in a new population. Then, it prints out the current population and draws the current best individual, as well as the speciation status, by calling `draw_current()`.

10. Once the loop exits, the results are printed, the truth table is checked, and the latest drawing is saved to an image file.

When running the program, the drawing containing the visualizations of the network and the speciation appears and updates itself at each generation, thereby creating an "animated" view of the status. The following figure contains four "snapshots" of the drawing that were captured during the run:

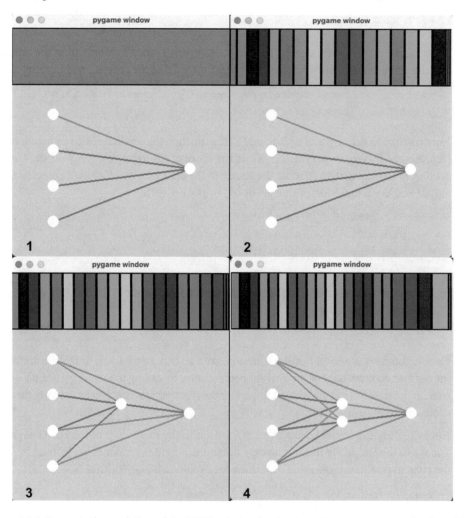

Figure 16.6: Stages in the evolution of the NEAT solution for the three-input even parity check problem

These snapshots demonstrate how the network starts with only the input and output layer nodes and a single species, then develops numerous species, followed by the addition of a single hidden layer node, and then a second one.

At the end of the run, the program saves one last snapshot as an image under the `images` folder. This looks as follows:

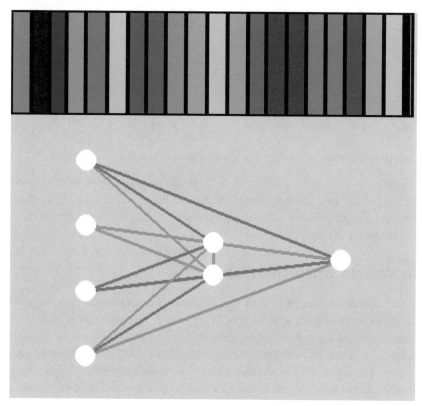

Figure 16.7: The final stage in the evolution of the NEAT solution
for the three-input even parity check problem

In the drawings, the white circles represent the nodes of the network, except for the top left circle, which is used to represent the *bias* values of the hidden and output layer nodes. The blue edges represent connections of positive weight (or a positive bias value), while the orange edges represent negative weights (or bias values). Unlike traditional MLPs, the networks created by the NEAT algorithm can have connections that "skip" a layer, such as the orange edge connecting the bottom input node directly to the output node, as well as intra-layer connections.

The printed output of the program indicates that the best network that was found was able to solve the problem:

```
best fitness = 7.9009068332812635
Number of nodes = 7
Checking the truth table:
input (0, 0, 0), expected output 0, got 0.050 -> 0
input (0, 0, 1), expected output 1, got 0.963 -> 1
input (0, 1, 0), expected output 1, got 0.933 -> 1
input (0, 1, 1), expected output 0, got 0.077 -> 0
input (1, 0, 0), expected output 1, got 0.902 -> 1
input (1, 0, 1), expected output 0, got 0.042 -> 0
input (1, 1, 0), expected output 0, got 0.029 -> 0
input (1, 1, 1), expected output 1, got 0.949 -> 1
```

As we can see, the best architecture that was found included a single hidden layer of two nodes.

In the next section, we will examine another biologically inspired, population-based algorithm. However, this algorithm deviates from using the familiar genetic operators of selection, crossover, and mutation, and instead utilizes a different set of rules to modify the population at each generation – welcome to the world of swarm behavior!

Particle swarm optimization

Particle swarm optimization (PSO) draws its inspiration from natural groupings of individual organisms, such as flocks of birds or schools of fish, generally referred to as *swarms*. The organisms interact within the swarm without central supervision, working together toward a common goal. This observed behavior gave rise to a computational method that can solve or optimize a given problem by using a group of candidate solutions, represented by *particles* analogous to organisms in a swarm. The particles move in the search space, looking for the best solution, and their movement is governed by simple rules that involve their position and *velocity* (directional speed).

The PSO algorithm is iterative, and in each iteration, every particle's position gets evaluated, and its best location so far, as well as the best location within the entire group of particles, are updated if necessary. Then, each particle's velocity is updated according to the following information:

- The particle's current speed and direction of movement – representing *inertia*
- The particle's best position found so far (local best) – representing *cognitive force*
- The entire group's best position found so far (global best) – representing *social force*

This is followed by an update to the particle's position, based on the newly calculated velocity.

This iterative process continues until some stopping condition, such as the *iterations limit*, is met. At this point, the group's current best position is taken as the solution by the algorithm.

This simple yet efficient process will be illustrated in detail in the next section, where we will go over a program that optimizes a function using the PSO algorithm.

PSO example – function optimization

For demonstration purposes, we will use the PSO algorithm to find the minimum location(s) of *Himmelblau's function*, a commonly used benchmark that we previously optimized using genetic algorithms in *Chapter 6, Optimizing Continuous Functions*. This function can be depicted as follows:

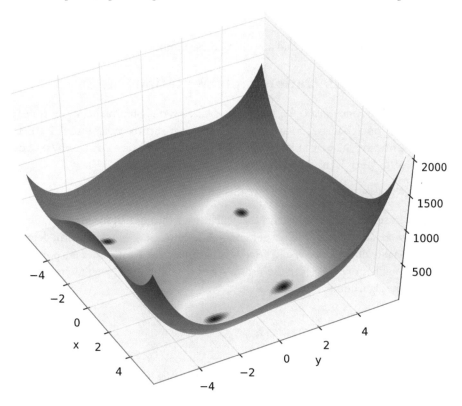

Figure 16.8: Himmelblau's function

Source: https://commons.wikimedia.org/wiki/
File:Himmelblau_function.svg

Image by Morn the Gorn.

As a reminder, the function can be mathematically expressed as follows:

$$f(x,y) = (x^2 + y - 11)^2 + (x + y^2 - 7)^2$$

It has four global minima, evaluating to 0, indicated by the blue areas in the plot. These are located at the following coordinates:

- *x=3.0, y=2.0*

- *x=−2.805118, y=3.131312*

- *x=−3.779310, y=−3.283186*

- *x=3.584458, y=−1.848126*

For our example, we will attempt to find any one of these minima.

Particle swarm optimization implementation

To locate a minimum of *Himmelblau's function* using particle swarm optimization, we've created a Python program called `04_pso_himmelblau.py` at `https://github.com/PacktPublishing/` `Hands-On-Genetic-Algorithms-with-Python-Second-Edition/blob/main/` `chapter_16/04_pso_himmelblau.py`.

The following steps describe the main parts of this program:

1. We start by setting various constants that will be used throughout the program. First, we have the *dimensionality* of the problem at hand – 2, in our case – which, in turn, determines the dimensionality of the *location* and *velocity* of each particle. Next comes the population size – the total number of particles in the swarm – and the number of generations, or iterations, of running the algorithm:

   ```
   DIMENSIONS = 2
   POPULATION_SIZE = 20
   MAX_GENERATIONS = 500
   ```

2. This is followed by several additional constants that affect how the particles are created and updated. We will see how they play their roles as we examine the rest of the program:

   ```
   MIN_START_POSITION, MAX_START_POSITION = -5, 5
   MIN_SPEED, MAX_SPEED = -3, 3
   MAX_LOCAL_UPDATE_FACTOR = MAX_GLOBAL_UPDATE_FACTOR = 2.0
   ```

3. Since our goal is to locate a minimum in *Himmelblau's function*, we need to define a single objective – that is, minimizing the fitness strategy:

   ```
   creator.create("FitnessMin", base.Fitness, weights=(-1.0,))
   ```

4. Now, we need to create the `Particle` class. Since this class represents a location in the continuous space, we could base it on an ordinary list of floats. However, here, we decided to use the `numpy` library's *N*-dimensional array (`ndarray`) since it lends itself to element-wise algebraic operations such as addition and multiplication, which will be needed when we update the particle's location. Besides the current location, the `Particle` class is given several additional attributes:

    - The `fitness` attribute, using the minimizing fitness we defined earlier.

    - The `speed` attribute, which is used to hold the current speed of the particle in each dimension. Although its initial value is `None`, `speed` will be populated with another `ndarray` later.

    - The `best` attribute, which represents the best location that's been recorded so far for this particular particle (*local best*).

    The resulting definition for the `Particle` class creator looks as follows:

    ```
    creator.create("Particle",
        np.ndarray,
        fitness=creator.FitnessMin,
        speed=None,
        best=None)
    ```

5. To help us construct an individual particle in the population, we need to define a helper function that will create and initialize a random particle. We will use the `numpy` library's `random.uniform()` function to randomly generate the location and speed arrays of the new particle, within the given boundaries:

    ```
    def createParticle():
        particle = creator.Particle(
            np.random.uniform(
                MIN_START_POSITION,
                MAX_START_POSITION,
                DIMENSIONS))

        particle.speed = np.random.uniform(
            MIN_SPEED,
            MAX_SPEED,
            DIMENSIONS)
        return particle
    ```

6. This function is used in the definition of the operator that creates a particle instance. This, in turn, is used by the population creation operator:

    ```
    toolbox.register("particleCreator", createParticle)
    toolbox.register("populationCreator",
                     tools.initRepeat,
    ```

```
list,
toolbox.particleCreator)
```

7. Next comes the method that serves as the heart of the algorithm, updateParticle(). This method is responsible for updating the location and speed of each particle in the population. The arguments of this function are a single particle in the population and the best currently recorded position.

    The method starts by creating two random update factors – one for the local update and the other for the global update – within the preset range. Then, it calculates two corresponding speed updates (local and global) and adds them to the current particle's speed.

    Note that all the values that are involved that are of the ndarray type are *two- dimensional* in our case, and the calculations are performed element-wise, one per dimension.

8. The updated particle speed is effectively a combination of the particle's original speed (representing *inertia*), the particle's best-known location (*cognitive force*), and the best-known location of the entire population (*social force*):

```python
def updateParticle(particle, best):
    localUpdateFactor = np.random.uniform(
        0,
        MAX_LOCAL_UPDATE_FACTOR,
        particle.size)
    globalUpdateFactor = np.random.uniform(
        0,
        MAX_GLOBAL_UPDATE_FACTOR,
        particle.size)
    localSpeedUpdate = localUpdateFactor *
        (particle.best - particle)
    globalSpeedUpdate = globalUpdateFactor * (best - particle)
    particle.speed = particle.speed +
        (localSpeedUpdate + lobalSpeedUpdate)
```

9. The updateParticle() method continues by making sure that the new speed does not exceed the preset limits and updates the location of the particles using the updated speed. As we mentioned previously, both location and speed are of the ndarray type and have separate components for each dimension:

```python
particle.speed = np.clip(particle.speed, MIN_SPEED, MAX_SPEED)
particle[:] = particle + particle.speed
```

10. Then, we must register the `updateParticle()` method as a toolbox operator that will be in the main loop later:

```
toolbox.register("update", updateParticle)
```

11. We still need to define the function to be optimized – *Himmelblau's function*, in our case – and register it as the fitness evaluation operator:

```
def himmelblau(particle):
    x = particle[0]
    y = particle[1]
    f = (x ** 2 + y - 11) ** 2 + (x + y ** 2 - 7) ** 2
    return f,  # return a tuple
toolbox.register("evaluate", himmelblau)
```

12. Now that we're finally at the `main()` method, we can start it by creating the population of particles:

```
population = toolbox.populationCreator(
    n=POPULATION_SIZE)
```

13. Before starting the algorithm's main loop, we need to create the `stats` object, to calculate the population's statistics, and the `logbook` object, to record the statistics at every iteration:

```
stats = tools.Statistics(lambda ind: ind.fitness.values)
stats.register("min", np.min)
stats.register("avg", np.mean)
logbook = tools.Logbook()
logbook.header = ["gen", "evals"] + stats.fields
```

14. The program's main loop contains an external loop that iterates over the generations/update cycles. Within each iteration, there are two secondary loops, each iterating over all the particles in the population. The first loop, which can be seen in the following code, evaluates each particle against the function to be optimized and updates the *local best* and the *global best* if necessary:

```
particle.fitness.values = toolbox.evaluate(particle)
# local best:
if (particle.best is None or particle.best.size == 0 or
    particle.best.fitness < particle.fitness):
    particle.best = creator.Particle(particle)
    particle.best.fitness.values = particle.fitness.values
# global best:
if (best is None or best.size == 0 or
    best.fitness < particle.fitness):
    best = creator.Particle(particle)
    best.fitness.values = particle.fitness.values
```

15. The second inner loop calls the `update` operator. As we saw previously, this operator updates the speed and the location of the particle using a combination of *inertia*, *cognitive force*, and *social force*:

```
toolbox.update(particle, best)
```

16. At the end of the outer loop, we record the statistics for the current generation and print them:

```
logbook.record(gen=generation,
               evals=len(population),
               **stats.compile(population))
print(logbook.stream)
```

17. Once the outer loop is done, we print the information for the best location that was recorded during the run. This is considered the solution that the algorithm has found for the problem at hand:

```
print("-- Best Particle = ", best)
print("-- Best Fitness = ", best.fitness.values[0])
```

By running this program, we get the following output:

```
gen evals min avg
0  20  8.74399 167.468
1  20  19.0871 357.577
2  20  32.4961 219.132
...
479 20 3.19693 316.08
480 20 0.00102484 322.134
481 20 3.32515 254.994
...
497 20 7.2162 412.189
498 20 6.87945 273.712
499 20 16.1034 272.385
-- Best Particle = [-3.77695478 -3.28649153]
-- Best Fitness = 0.0010248367255068806
```

These results indicate that the algorithm was able to locate one of the minima, around x=−3.77 and y=−3.28. Looking at the stats we recorded along the way, we can see that the best result was achieved at generation 480. It is also evident that the particles move around quite a bit and, during the run, oscillate about the best result.

To find the other minimum locations, you can rerun the algorithm with a different random seed. You can also penalize the solutions in the areas around the previously found minima, just like we did with *Simionescu's function* in *Chapter 6, Optimizing Continuous Functions*. Another approach could be using multiple simultaneous swarms to locate several minima in the same run – you are encouraged to try this on your own (see the *Further reading* section for more information).

In the next section, we will briefly review several more members of the extended evolutionary computation family.

Other related techniques

Besides the techniques we have covered so far, numerous other problem-solving and optimization techniques draw their inspiration from the Darwinian evolution theory, as well as from various biological systems and behaviors. The following subsections briefly describe several more of these techniques.

Evolution strategies

Evolution strategies (**ES**) are a kind of genetic algorithm that emphasizes *mutation* rather than *crossover* as the evolutionary facilitator. The mutation is adaptive, and its strength is learned over the generations. The selection operator in ES is always based on *rank* rather than on actual fitness values. A simple version of this technique is called *(1 + 1)*. It includes only two individuals – a parent and its mutated offspring. The best of them continue to be the parent of the next mutated offspring. In the more general case, called *(1 + λ)*, there is one parent and λ mutated offspring, and the best of the offspring continues to be the parent of the next λ offspring. Some newer variations of the algorithm include more than one parent, as well as a *crossover* operator.

Differential evolution

Differential evolution (**DE**) is a specialized variant of genetic algorithms that's used to optimize real-valued functions. DE differs from genetic algorithms in the following aspects:

- The DE population is always represented as a collection of real-valued vectors.

- Instead of replacing the entire current generation with a new generation, DE keeps iterating over the population, modifying one individual at a time, or keeping the original individual if it's better than its modified version.

- The traditional *crossover* and *mutation* operators are replaced by specialized ones, thereby modifying the value of the current individual using the values of three other individuals that are chosen at random.

Ant colony optimization

Ant colony optimization (**ACO**) algorithms are inspired by the way certain species of ants locate food. The ants start by wandering randomly, and when any of them locates food, they go back to their colony while depositing pheromones along the way, marking the path for other ants. Other ants finding food at the same location will reinforce the trail by depositing their own pheromones. The pheromone marks fade away over time, giving the shorter paths and the paths that are traveled more often an advantage.

ACO algorithms use artificial ants that move about in the search space looking for the location of the best solutions. The "ants" keep track of their locations and the candidate solutions they have found along the way. This information is used by the ants of the subsequent iterations so that they can find better solutions. These algorithms are often combined with the *local search* method, which is activated after locating an area of interest.

Artificial immune systems

Artificial immune systems (**AIS**) draw their inspiration from the characteristics of adaptive immune systems found in mammals. These systems are capable of identifying and learning new threats, as well as applying the acquired knowledge and responding faster the next time a similar threat is detected.

Recent AIS can be used in various machine learning and optimization tasks, and generally belong to one of the following three subfields:

- **Clonal selection**: This involves imitating the process by which the immune system selects the best cell to recognize and eliminate an antigen that enters the body. The cell is chosen out of a pool of pre-existing cells with varying specificities, and once chosen, it is cloned to create a population of cells that eliminates the invading antigen. This paradigm is typically applied to optimization and pattern recognition tasks.

- **Negative selection**: This follows a process that identifies and deletes cells that may attack self-tissues. These algorithms are typically used in anomaly detection tasks, where normal patterns are used to "negatively" train filters that will then be able to detect anomalous patterns.

- **Immune network algorithms**: This is inspired by the theory that suggests that the immune system is regulated using special types of antibodies that bind to other antibodies. In this type of algorithm, antibodies represent nodes in a network and the learning process involves creating or removing edges between the nodes, resulting in an evolving network graph structure. These algorithms are typically used in non-supervised machine learning tasks, as well as in the fields of control and optimization.

Artificial life

Rather than being a branch of evolutionary computation, **artificial life** (**ALife**) is a broader field that involves systems and processes that imitate natural life in different ways, such as computer simulations and robotic systems.

Evolutionary computation can be viewed as an application of ALife, where the population seeking to optimize a certain fitness function is a metaphor for organisms searching for food. The niching and sharing mechanisms, which we described in *Chapter 2, Understanding the Key Components of Genetic Algorithms*, draw directly from the food metaphor.

The main branches of ALife are as follows:

- **Soft**: Represents software-based (digital) simulation
- **Hard**: Represents hardware-based (physical) robotics
- **Wet**: Represents biochemical-based manipulation or synthetic biology

ALife can also be viewed as the bottom-up counterpart to artificial intelligence since ALife typically builds on the biological environment, mechanisms, and structures rather than high-level cognition.

Summary

In this chapter, you were introduced to the extended family of *evolutionary computation* and some of the common characteristics of its members. Then, we used *genetic programming* – a special case of genetic algorithms – to implement the *even parity check* task using Boolean logic building blocks.

Next, we created a neural network implementation of the same even parity check task by utilizing the NEAT technique.

This was followed by creating a program that utilized the *particle swarm optimization* technique to optimize *Himmelblau's function*.

We concluded this chapter with a brief overview of several other related problem-solving techniques.

Now that this book has come to its end, I wanted to thank you for taking this journey with me while going through the various aspects and use cases of genetic algorithms and evolutionary computation. I hope that you found this book interesting as well as thought-provoking. As this book demonstrated, genetic algorithms and their related techniques can be applied to a plethora of tasks in virtually any computation and engineering field, including – very likely – the ones you are currently involved with. Remember, all that is required for the genetic algorithm to start crunching a problem is a way to represent a solution and a way to evaluate a solution – or compare two solutions. Since this is the age of artificial intelligence and cloud computing, you will find that genetic algorithms lend themselves well to both and can be a powerful tool in your arsenal when you're approaching a new challenge.

Further reading

For more information, please refer to the following resources:

- *Genetic Programming: bio-inspired machine learning*: `http://geneticprogramming.com/tutorial/`
- *Artificial Intelligence for Big Data*, by Manish Kumar and Anand Deshpande, May 21, 2018
- *Hands-On Neuroevolution with Python*, by Iaroslav Omelianenko, December 24, 2019
- *Multimodal optimization using particle swarm optimization algorithms*: CEC 2015 competition on single objective multi-niche optimization: `https://ieeexplore.ieee.org/document/7257009`

Index

Packtpub.com

Subscribe to our online digital library for full access to over 7,000 books and videos, as well as industry leading tools to help you plan your personal development and advance your career. For more information, please visit our website.

Why subscribe?

- Spend less time learning and more time coding with practical eBooks and Videos from over 4,000 industry professionals

- Improve your learning with Skill Plans built especially for you

- Get a free eBook or video every month

- Fully searchable for easy access to vital information

- Copy and paste, print, and bookmark content

Did you know that Packt offers eBook versions of every book published, with PDF and ePub files available? You can upgrade to the eBook version at packtpub.com and as a print book customer, you are entitled to a discount on the eBook copy. Get in touch with us at customercare@packtpub.com for more details.

At www.packtpub.com, you can also read a collection of free technical articles, sign up for a range of free newsletters, and receive exclusive discounts and offers on Packt books and eBooks.

Other Books You May Enjoy

If you enjoyed this book, you may be interested in these other books by Packt:

OpenAI API Cookbook

Henry Habib

ISBN: 978-1-80512-135-0

- Grasp the fundamentals of the OpenAI API
- Navigate the API's capabilities and limitations of the API
- Set up the OpenAI API with step-by-step instructions, from obtaining your API key to making your first call
- Explore advanced features such as system messages, fine-tuning, and the effects of different parameters
- Integrate the OpenAI API into existing applications and workflows to enhance their functionality with AI
- Design and build applications that fully harness the power of ChatGPT

Unlocking the Secrets of Prompt Engineering

Gilbert Mizrahi

ISBN: 978-1-83508-383-3

- Explore the different types of prompts, their strengths, and weaknesses
- Understand the AI agent's knowledge and mental model
- Enhance your creative writing with AI insights for fiction and poetry
- Develop advanced skills in AI chatbot creation and deployment
- Discover how AI will transform industries such as education, legal, and others
- Integrate LLMs with various tools to boost productivity
- Understand AI ethics and best practices, and navigate limitations effectively
- Experiment and optimize AI techniques for best results

Packt is searching for authors like you

If you're interested in becoming an author for Packt, please visit authors.packtpub.com and apply today. We have worked with thousands of developers and tech professionals, just like you, to help them share their insight with the global tech community. You can make a general application, apply for a specific hot topic that we are recruiting an author for, or submit your own idea.

Share Your Thoughts

Now you've finished *Hands-On Genetic Algorithms with Python, Second Edition*, we'd love to hear your thoughts! Scan the QR code below to go straight to the Amazon review page for this book and share your feedback or leave a review on the site that you purchased it from.

https://packt.link/r/1-805-12379-3

Your review is important to us and the tech community and will help us make sure we're delivering excellent quality content.

Download a free PDF copy of this book

Thanks for purchasing this book!

Do you like to read on the go but are unable to carry your print books everywhere?

Is your eBook purchase not compatible with the device of your choice?

Don't worry, now with every Packt book you get a DRM-free PDF version of that book at no cost.

Read anywhere, any place, on any device. Search, copy, and paste code from your favorite technical books directly into your application.

The perks don't stop there, you can get exclusive access to discounts, newsletters, and great free content in your inbox daily

Follow these simple steps to get the benefits:

1. Scan the QR code or visit the link below

https://packt.link/free-ebook/978-1-80512-379-8

2. Submit your proof of purchase
3. That's it! We'll send your free PDF and other benefits to your email directly

Made in the USA
Thornton, CO
07/05/24 13:54:02

161d24a0-ee50-4801-b73e-b59f8607a9c8R01